JOSHUA KONSTANTINOS

Sleeping On a Volcano

The Worldwide Demographic Upheaval and the Economic and Geopolitical Implications

Copyright © 2019 by Joshua Konstantinos

All rights reserved. No part of this publication may be reproduced, stored or transmitted in any form or by any means, electronic, mechanical, photocopying, recording, scanning, or otherwise without written permission from the publisher. It is illegal to copy this book, post it to a website, or distribute it by any other means without permission.

First edition

This book was professionally typeset on Reedsy. Find out more at reedsy.com

To my loving wife, Nurcan, who has been understanding in allowing me to devote time to this project, immeasurably helpful in editing, and has threatened me with physical violence if I do not include her in the dedications.

And to the rest of my family who has been forced to listen to me rant on this topic for years – now you'll have to read a whole about it book too.

"I am told there is no danger because there are no riots. I am told that because there is no visible disorder on the surface of society that there is no revolution at hand. Gentleman, permit me to say that I believe you are mistaken...We are sleeping on a volcano. Do you not see the Earth trembles anew? A wind of revolution blows, a storm is on the horizon."

Alexis de Tocqueville, addressing the French parliament months before the revolutions of 1848

Contents

I   Sleeping on a Volcano

Sleeping on A Volcano   3
  Predicting the Crisis   8
  The Larger Crisis   11
      15

II   Pressure Building

The Demographic Upheaval and Its Economic Impact   19
  The Scale of the Demographic Upheaval   21
    The Optimistic Nature of the Population Data   38
  Economic Impact of the Demographic Decline   40
    End of the Demographic Dividend   42
    Rising Interest Rates   45
    Fall In Demand for Bonds and Other Assets   46
    Surge in Costs   48
    Economic Shock   51
      52

The Rise and Fall of the Bretton Woods System   53
  The Bretton Woods Agreement   54
    The Triffin Dilemma   59
    Collapse   60
  The Imbalanced Global Economy   61

Global Trade	62
Free-Floating Currencies and Debt.	70
Free-Floating Currencies and Inflation	75
The Impact of Free-Floating Currencies	78
	78

III   Unsustainible Accumulation of Debt

The Accumulation of Debt and Quantitative Easing	81
The Impact of a Sovereign Debt Crisis	81
Why We Haven't Seen A Crisis Yet	89
Quantitative Easing	91
The Economic Distortion of Artificially low Interest Rates	96
Sovereign Debt by Country	102
Japan	103
The Plaza Accords	103
Japanese Sovereign Debt	106
Consensus of Unsustainablity	112
	120
Italy and Europe	121
The European Union	121
The Italian Crisis	132
Italian Fundamentals	135
The Threat to Europe and the Rest Of the World	140
China	143
The Rise of China	143
Debt Binge in China	146
Chinese Debts in Foreign Currency	150
Chinese Asset Bubble	152
The United States	155
The United States of Debt	157
Debt Rollover	163
	166

	166
	166
	166

IV Financial Crises and The Sovereign Debt Crisis

The Sovereign Debt Crisis	169
The Money Supply, Prices, and Resource Allocation	171
Changes to the Money Supply	171
Fractional Reserve Banking	174
Volcanic Eruption	181
The Risk from Debt	182
Default and Deflation	185
Austerity	186
Monetization and Inflation	187
What is Now Inevitable	189
Debt Forgiveness	191
	192
	192
Notes	193

I

Sleeping on a Volcano

1

Sleeping on A Volcano

In January of 1848, although only months before a wave of revolutions would sweep across the continent, Europe was seemingly at peace. Alexis de Tocqueville, the prescient author of *Democracy in America*, addressed the French parliament warning them of the imminent upheaval saying:

> *I am told there is no danger because there are no riots. I am told that because there is no visible disorder on the surface of society that there is no revolution at hand. Gentleman, permit me to say that I believe you are mistaken...We are sleeping on a volcano. Do you not see the Earth trembles anew? A wind of revolution blows, a storm is on the horizon.*

Today our world is again sleeping on a volcano - poised for economic and geopolitical upheaval. The economic and geopolitical pressures, which have been building for decades, will inevitably erupt - triggered by the aging of the world.

A*ging* sounds like such a gradual process; that is not how the demographic transition from young to old will impact the world. The Baby Boomer generation will retire quite abruptly. From a historical perspective, the generation will go from being in their peak earning years to retired in the blink of an eye. And because fertility rates have collapsed around the world since the 1970s, in most of the developed world the number of working-age adults will plummet with the Baby Boomer retirement.

Tax revenues will fall sharply in many nations while costs for pensions and healthcare rise at the same time. This economic difficulty will be made many times worse because of the current state of the global economy. Following the Great Recession, the nations of the world took on a massive amount of debt to bailout their financial systems. And for the last decade, central banks around the world have used *unconventional monetary policy* to lower long-term interest rates to ultra-low and even negative rates. This has incentivized even more debt accumulation. Now, with the transition of the Baby Boomers from taxpayers to pensioners, the solvency not only of banks but of nations will be questioned in a sovereign debt crisis.

This is a crisis that has been in the making for more than half a century, and there is more to the story than the shifting demographics. As we will see, several other factors have also contributed. However, to understand the peril facing the world, we should begin a little closer to the present day - with the Great Recession.

* * *

In late 2006, a massive real estate bubble in the United States that had been created over the previous decade began to stall and then quickly deflate.

As housing prices fell, more and more people, especially low-credit or *subprime* borrowers, defaulted on their mortgages. The first sign of an actual *crisis,* however, did not occur until August 9th 2007, when the French bank *BNP Paribas* was forced to freeze the money in their investor's subprime funds. This shock exacerbated the problems in the banking system considerably. Banks became reluctant to lend to other banks, unsure if they would need the cash for themselves – or if their counter-parties would be able to repay them. Who really knew which banks had invested heavily in subprime mortgages and were already insolvent?

Despite actions by the central banks to calm markets and provide liquidity to the financial system, the crisis continued to worsen as the housing market maintained its free fall. On March 14th 2008, the Federal Reserve held an emergency weekend meeting – it's first in over thirty years. The central bank announced they would guarantee the bad loans of the now insolvent Bear Stearns, bailing out the investment bank. In addition, Fannie Mae and Freddie Mac, the government-sponsored enterprises, would purchase another two hundred billion dollars of subprime mortgage debt. However, the crisis continued. On July 11th, IndyMac Bank in California failed - the fourth largest bank failure in U.S. history. Depositors waited in line to withdraw their deposits, which at that point were only insured by the FDIC up to $100,000.00. On September 7th, the Treasury Department took Fannie Mae and Freddie Mac into conservatorship - meaning that they would now be run directly by the federal government.

Everything that had happened up till this point was fairly typical of a recession. What made this *The Great Recession* was the collapse of Lehman Brothers. Like Bear Stearns, Dick Fuld's Lehman Brothers had invested heavily in mortgage debt and was now in dire straits. On Saturday, Sept. 13th, the Federal Reserve allowed the big banks to pledge new types of collateral in return for overnight capital. Dick Fuld soon learned that this exemption was being extended to everyone *but* Lehman[1]. On Sunday September 14th, officials from Federal Reserve and the U.S. Treasury Department met to find a buyer for Lehman

Brothers. They needed a solution before the Asian markets opened and Barclays and Bank of America were in talks with the government to buy Lehman Brothers. However, former Goldman Sachs CEO Henry Paulson was now the head of the Treasury Department. Paulson and other members of the Treasury and the Fed were concerned about the moral hazard of bailing out Lehman. Lehman Brothers had been egregious in its risky investment into subprime debt – and had benefited from it tremendously before the crisis. And, reportedly, there was no love lost between Paulson and Fuld[2]. Paulson refused to sweeten the deal with public funds and both Barclays and Bank of America walked away from the deal. Ultimately, the decision was made to allow Lehman Brothers to go bankrupt. The world soon realized that Lehman Brothers had been *too big to fail* - that is too big to *allow* to fail as its failure would take the rest of the economy with it.

With the collapse of Lehman Brothers, the banking system was now unsure if the Federal Government would be there as the lender of last resort. Interbank lending froze. Many on Wall Street and others following the situation began to pull their money out of banks in a replay of the Great Depression. The Hendersonville Times-News reported U.S. senator Richard Burr's reaction after being briefed on the situation:

> *On Friday night, I called my wife and I said, 'Brooke, I am not coming home this weekend. I will call you on Monday. Tonight, I want you to go to the ATM machine, and I want you to draw out everything it will let you take. And I want you to tomorrow, and I want you to go Sunday'. I was convinced on Friday night that if you put a plastic card in an ATM machine the last thing you were going to get was cash*[3].

The senator was not alone. Days after the collapse of Lehman Brothers, investors withdrew a record $144.5 billion from their money market accounts vs the typical 7 billion[4]. Large businesses were having difficultly having money transferred to meet payroll obligations[5]. The entire banking system had seized-up and if this had continued, businesses would not have been able to get money to fund their day-to-day operations[6]. Businesses would have been unable to make payrolls, and depositors would have been unable to withdraw cash from ATMs. The economy would have collapsed.

In the days after the Lehman bankruptcy the need for drastic action became apparent. Treasury Secretary Henry Paulson reportedly literally fell to his knees begging Speaker of the House of Representatives, Nancy Pelosi, to support a bailout of the financial system[7]. Ben Bernanke informed lawmakers that the banking system would cease to exist if they did not act quickly. Paulson warned they would need to plan for martial law. And that if they did not bailout the economy, Congress would need a plan to feed the country if banking and commerce collapsed.

In an in-depth review of the collapse of Lehman Brothers, Stephen Foley, writing in the *Independent,* quotes Alan Blinder, the former Federal Reserve board member and professor at Princeton University, on the impact of the Lehman collapse:

> *People argue that if it wasn't Lehman Brothers it would have been something else," ...."I don't buy that. I don't mean everything would have been great if we had bailed out Lehman. We were in a financial crisis before Lehman. But it had a shock value that just caused everything to fall off a cliff. If you look at data on almost anything – consumer spending, investment spending, car sales, employment – it just drops off the table at Lehman Brothers and I don't think we needed to have that*[8].

The seven hundred billion dollar taxpayer bailout that congress passed helped prevent the collapse of the global economy. However, all told, the Federal Reserve committed 7.7 trillion dollars to stabilizing the global economy in the six months after Lehman collapsed[9]. Nevertheless, much of the damage had already been done. And as we will see, the crisis itself was only a part of a larger systemic problem; and the beginning of an acceleration towards an even larger crisis.

Predicting the Crisis

Most people would say that the 2008 crisis caught the world by surprise. *Maybe the sudden and unexpected fall in housing prices should have been foreseen*, they argue, *but only in retrospect were the signs clear.* This is the argument that former Federal Reserve chairman Alan Greenspan has made. "Everybody missed it — academia, the Federal Reserve, all regulators[10]" he said in 2010. In his view, central banks, government watchdogs, and banking and finance in general were all caught off guard. Like investors and the major financial institutions they were blindsided – thrust into the crisis without warning.

But this is simply not the case.

In fact, William White, the chief economist at arguably the most prestigious and oldest international financial institution, the Bank for International Settlements (BIS), warned the leaders of global finance for years leading up to the crisis. White specifically highlighted the bubble in the U.S. housing market and the risk of contagion from subprime mortgage loans. The German paper *Der Spiegel* reported in 2009 how the world ignored his warnings:

> *Now White has been proved right — to an almost apocalyptical degree. And yet gloating is the last thing on his mind. He, the chief economist at the central bank for central banks, predicted the disaster, and yet not even his own clientele was willing to believe*

> him. It was probably the biggest failure of the world's central bankers since the founding of the BIS[Bank for International Settlements] in 1930. They knew everything and did nothing. Their gigantic machinery of analysis kept spitting out new scenarios of doom, but they might as well have been transmitted directly into space.
>
> For years, the regulators of the global money supply ignored the advice of their top experts, probably because it would require them to do something unheard of, namely embark on a fundamental change in direction[11].

Likewise, the chief economist at the International Monetary Fund (IMF), Ken Rogoff, warned of the crisis as early as 2001 in a series of papers predicting the collapse of the housing market. "It got no interest from anybody" he later recounted[12].

Micheal Burry of Scion Capital - famously portrayed in the movie *the Big Short* - who correctly predicted and profited from the subprime mortgage crash, opined in the New York Times:

> ..the signs were all there in 2005, when a bursting of the bubble would have had far less dire consequences, and when the government could have acted to minimize the fallout.
>
> Instead, our leaders in Washington either willfully or ignorantly aided and abetted the bubble. And even when the full extent of the financial crisis became painfully clear early in 2007, the Federal Reserve chairman, the Treasury secretary, the President and senior members of Congress repeatedly underestimated the severity of the problem, ultimately leaving themselves with only one policy tool — the epic and unfair taxpayer-financed bailouts. Now, in exchange for that extra year or two of consumer bliss we

all enjoyed, our children and our children's children will suffer terrible financial consequences[13].

White and the others who foresaw the crash did not have some super sophisticated model that allowed them predict the crisis. Nor were they savants who saw things that others simply couldn't see. Indeed, even the Mortgage Insurance Companies of America (MICA), wrote to the Federal Reserve warning them of their concern for the U.S. real estate market – citing a study entitled *This Powder Keg Is Going to Blow*[14]. This study and others clearly warned there was were significant problems brewing in the housing market.

The bubble in the U.S. housing market was obvious; the issue was that the people in finance participating in the bubble had a strong financial incentive not to see the bubble. And the national centrals banks of the world had similar political incentives not to see the impending crisis. As they say – *It is difficult to get a man to understand something when his salary depends upon his not understanding it.* However, simply looking at a graph of home prices shows the bubble in housing. The imminent crisis was clearly visible – a major and unwarranted deviation from decades of data that made earlier housing bubbles look insignificant.

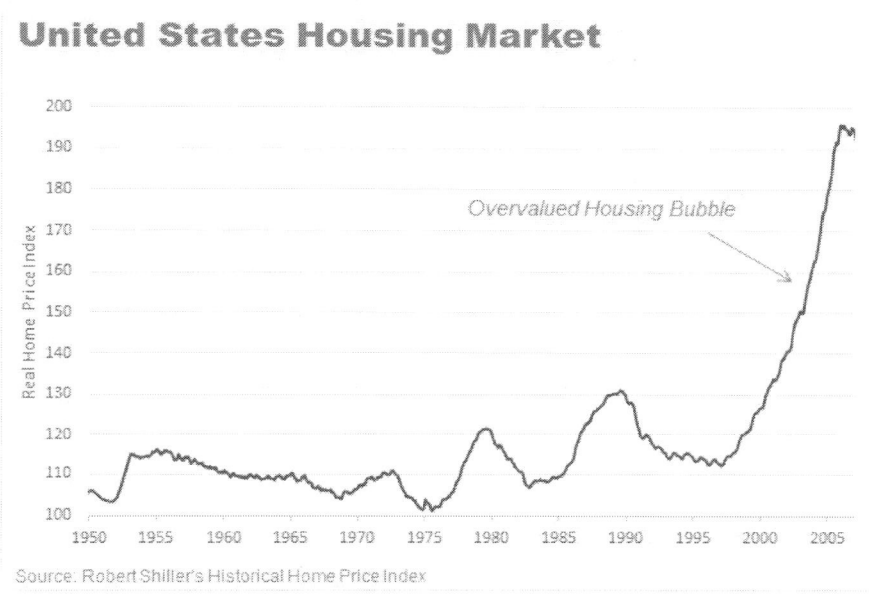

The Larger Crisis

In 2018, just before the World Economic Forum in Davos Switzerland, Dr. William White warned the world of another approaching crisis, a crisis which makes the world today look "worse than 2008[15]" Moreover, this future catastrophe has been in the making for much longer then the Great Recession's housing bubble.

White explained that the dramatic demographic changes the world has seen in the past half-century have had enormous economic consequences. The massive increase in labor with the worldwide Baby Boom and the integration of China and Eastern Europe into the global economy was the largest positive labor shock in human history. The increase in the supply of labor led to stagnate wages in developed countries – but also rapid GDP growth around the world. Sharply falling fertility rates – particularly in Asia where the fall was most extreme – provided a so called *demographic dividend* to the world. As the Boomer generation had substantially less children than their parents, this reduced the dependency ratio (the ratio between working-age adults and the old and young which depend on them) and allowed more women to enter the workforce. Additionally, as the Boomers saved more and more for retirement all the money they saved helped push down real interest rates lower for decades.

The economic boost from growing populations, an age structure which increased the proportion of working-age adults, the rise of globalization, and low interest rates, created decades of strong economic growth. This environment of high GDP growth and falling real interest rates allowed the world to build up massive amounts of debt, especially debt issued by sovereign nations, known as *sovereign debt*.

This debt build-up has been going on for decades. However, since the 2008 financial crisis, the process has accelerated. After the the Great Recession, central banks manipulated global markets with ultra-low interest rates by using so called *unconventional monetary policy*. This prevented a immediate debt crisis at the time, but worsened the inevitable reckoning by distorting the global economy. Creating *zombie companies* – companies which do not make enough profit to pay the interest on their existing debt but who must continually to borrow to stay afloat and so are dependent on low interest rates.

With the imminent retirement of the Baby Boomers, central banks will not be able to maintain low interest rates forever. As White warned at Davos:

> *Everything could now go into reverse: the baby boomers are gone; China's working age population is falling; and zombie companies are going to be forced out of business at last as borrowing costs rise...*

Central banks are now caught in a "debt trap". They cannot hold rates near zero as inflation pressures build, but they cannot easily raise rates either because it risks blowing up the system. I am afraid that at some point this is going to be resolved with a lot of debt defaults. And what did we do with the demographic dividend? We wasted it... It is frankly scary[16]

And while White may be one of the most prominent and well respected voices issuing warning of this future crisis, he is far from the only voice. Michael Burry when recently interviewed for the article, *Michael Burry, Real-Life Market Genius From The Big Short, Thinks Another Financial Crisis Is Looming*, had this to say:

It seems the world is headed toward negative real interest rates on a global scale. This is toxic. Interest rates are used to price risk, and so in the current environment, the risk-pricing mechanism is broken. That is not healthy for an economy. We are building up terrific stresses in the system, and any fault lines there will certainly harm the outlook.

...The public sector has really stepped up as a consumer of debt. The Federal Reserve's balance sheet is leveraged 77:1. Like I said, the absurdity, it just befuddles me.

Reuters reported in May 2019 the warning of bankers assembled in Stockholm for the annual meeting of the International Capital Markets Association:

Heavy reliance on debt financing and slow economic growth are leading to the creation of debt bubbles which risk destabilizing the entire financial system should a major shock occur, bond bankers and investors warned at a conference on Thursday.

Leverage in the system has increased some 50%-60% since the financial crisis a decade ago, with debt now worth some 230% of economic output globally according to IMF estimates, said Frank Czichowski, treasurer at KfW.

"There is no banking system in the world that can cope with the flood coming out of this," he said.

Affordability is not an issue while rates remain low but a rise in borrowing costs could create major problems, Axa's Stoter said: "The question is, with the debt level where it is, can central banks ever afford to let interest rates go back up because it will lead to a major bankruptcy wave.[17]"

Over the last five decades, governments around the world have taken advantage of the massive size of the Baby Boomer generation to accumulate massive debts. And after the Great Recession, nations took on massive amounts of debt to bailout their financial systems. Since then, central banks have been forced to use *quantitative easing* (QE) to manipulate long-term interest rates lower throughout the global economy. This has delayed the popping of the sovereign debt bubble that has been building for fifty years - but it has also interfered with resource allocation on a massive scale. Not only governments, but corporations and households have become dependent on borrowing at ultra-low interest rates.

As with the 2008 housing crash, many in finance and politics have a strong incentive not to see an approaching crisis; but anyone who takes an unbiased look the available information can clearly see the inevitable outcome. The world is headed for a crisis that has the potential to make 2008 pale in comparison - and there is no entity who can bailout national governments on this scale. Decades of misallocation of resources, ultra-low interest rates, and massive accumulation of debts by households, corporations, and governments will be brought to an abrupt end with the retirement of the Baby Boomer generation.

It is difficult to make the case for an inevitable crisis. But while this argument may seem extreme, as we will see throughout this book, this view is not on the fringe – but supported by experts from leading institutions who have been warning for decades of a debt crisis. In many cases the same respected experts, such as the former head of the IMF Ken Rogoff and the former head of the BIS William White, who warned the world of the 2008 housing crisis.

In 2018, the IMF said that "large challenges loom for the global economy to prevent a second Great Depression" and that "With global debt levels well above those at the time of the last crash in 2008, the risk remains that unregulated parts of the financial system could trigger a global panic.[18]" It is in this economic environment that the Baby Boomer generation is about to retire en mass. And this demographic upheaval will, as experts have been predicting for decades, have a major negative economic impact and put tremendous stress on the global economy and national budgets.

To understand this future crisis we need to understand what brought the world to this point. Before looking towards the future we must grasp the demographic, economic, and geopolitical changes since the end of World War II, and how they have set the world on the path to this crisis.

II

Pressure Building

Fifty Years of Demographic and Economic Pressure

2

The Demographic Upheaval and Its Economic Impact

In 1968, Paul Ehrlich published *The Population Bomb;* opening with the line "The battle to feed all of humanity is over." He argued that the increase in fertility rates and the accompanying surge in population that was the *Baby Boom* meant that "hundreds of millions of people are going to starve to death." The enormous increase in world population did not result in worldwide hunger as Ehrlich predicted, largely because of the work of Norman Borlaug – *the man who saved a billion lives*[19].

Norman Borlaug

Norman Borlaug was the father of the Green Revolution in agriculture. His development of high yield and disease resistant wheat strains prevented worldwide hunger and allowed the planet to feed the massive increase in population. In acknowledgement of the profound impact of his work, Borlaug was awarded the 1970 Nobel Peace Prize, the Presidential Medal of Freedom, and the Congressional Gold Medal.

Without the increase in crop yields the worldwide population explosion that was the *Baby Boomer* generation would have undoubtedly led to starvation in some areas of the world. Yet, by the time that Paul Ehrlich published his influential work, an even more unprecedented demographic change had already begun – a collapse in fertility rates, unlike anything ever seen before in human history[20]. The development of birth control, the rapid urbanization of the world, declining infant mortality rates, and a combination of other factors[21], meant that by 1975 much of the developed world's fertility rates had fallen below the minimum population replacement rate. The future would not be one of overpopulation; generations following the *Baby Boomers* ended centuries of exponential population growth and are either similar in size or, as in most countries, substantially smaller.

This fall in fertility rates has been much more consequential than most people realize. As we will see, major events from the rise of economic giants in Asia to stagnate wages in the developed world can be directly attributed to falling fertility rates. But the real consequences will arise when the Baby Boomer generation retires. But before we look at the consequences of the Baby Boomer retirement, we much understand the scale of the demographic upheaval the world has seen since the end of World War Two.

* * *

The Scale of the Demographic Upheaval

Most people do not realize the magnitude of the post-war Baby Boom or the fall in fertility rates that followed it. When fertility rates fall below 2.1 births per woman (the point known as the minimum population replacement level), each preceding generation will be smaller than the last. Many nations in the developed world have been at almost half of the minimum replacement rate for decades. Meaning that the generations after the Baby Boom were only just over half the size of the previous generations in many countries.

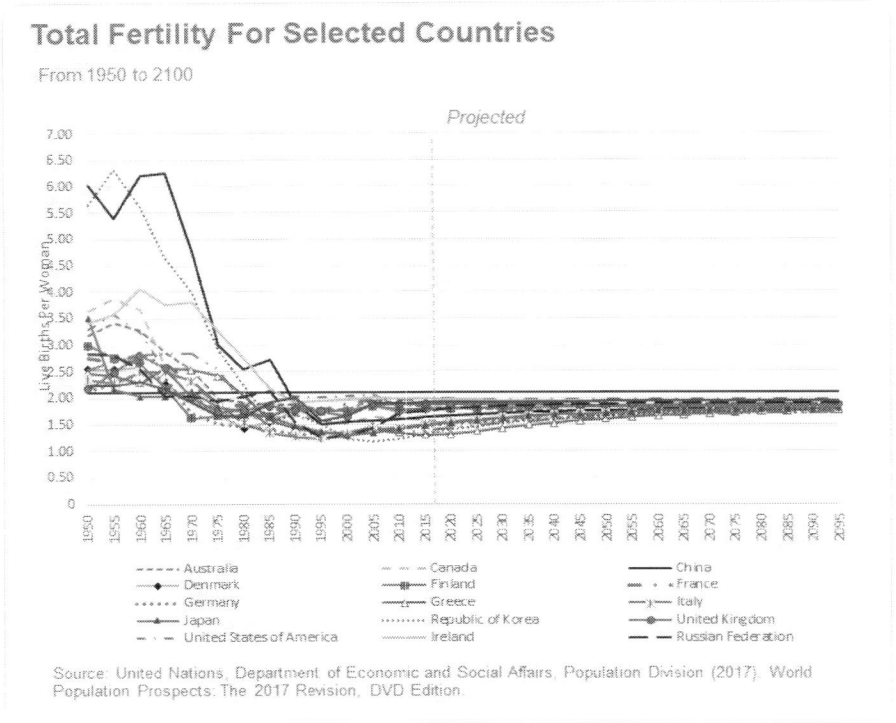

When looking at fertility rates, it's important to understand their impact on population levels. If a country has a fertility rate of 6, as China did around 1960, then the next generation will be **three times** larger than the generation before it. One billion people grows to three billion in a generation. Then nine billion the next, then 27 billion. The growth rate is exponential. A fertility rate of 2.1 keeps the population stable, but when it drops below 2.1 the population begins to fall exponentially. Rates have been far below 2.1 for several generations now in most countries – and rates have continued to fall even lower. In 2017, Singapore had the world's lowest fertility rate – measured at 0.83[22]. Hong Kong's more typical 1.19 fertility rate means that the next generation will be forty percent smaller than the previous generation.

In this case, one billion turns into roughly six hundred million, then three hundred and fifty million, then two hundred and ten million. In three generations a nation with a 1.19 fertility rate will see the size of their generations fall almost eighty percent.

The graph below illustrates the impact of fertility rates on the size of the following generations. Starting with a hypothetical population of one million people, you can see that a sustained fertility rate of 3 creates a 440% increase to 4.4 million in four generations, while a rate of 1.2 leads to a generation of only one hundred and six thousand – an almost 90% decrease in four generations.

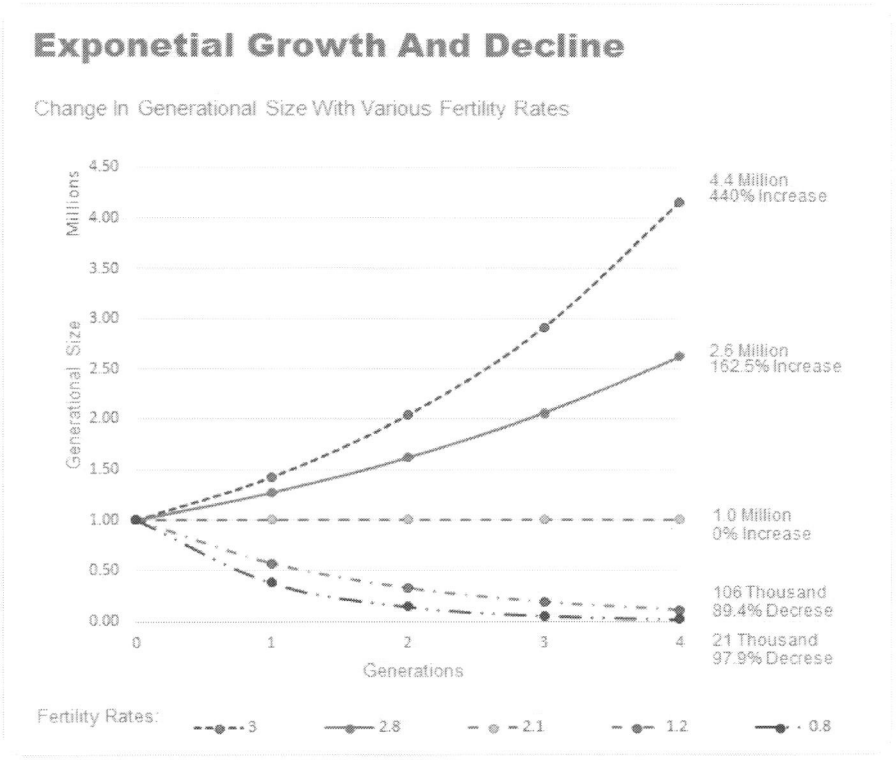

Looking more closely at the fertility rates in the G7 nations we can see two important developments that have had a major impact already. Firstly, the early fall in the Japanese fertility rates. Because of this early drop, they are one of the countries that has aged the most already. Secondly, we can see that the Untied States fertility rate hovered very close to the minimum population replacement rate of 2.1 for decades. Additionally, the U.S. has seen a substantial amount of immigration, so while the United States is ending centuries of growth, the population is is fairly stable in the short term and not actually declining as has already begun in Japan.

THE DEMOGRAPHIC UPHEAVAL AND ITS ECONOMIC IMPACT

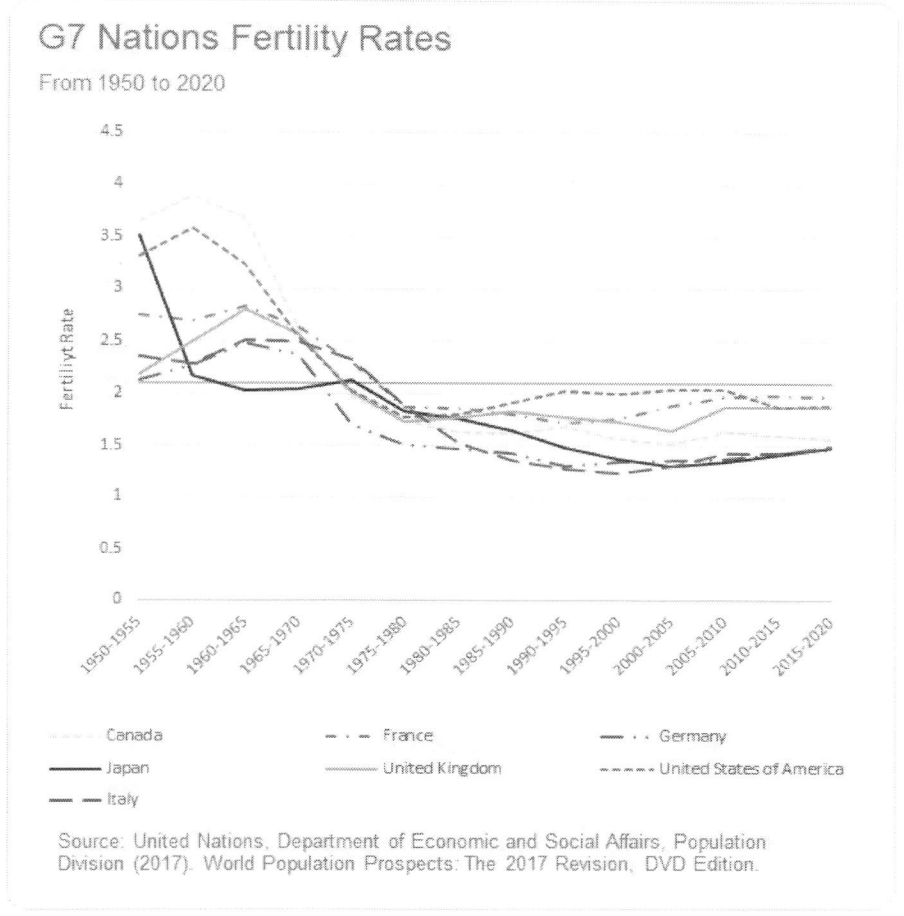

While fertility rates have fallen, life expectancy has continued to rise. In 1950 the average lifespan ranged from sixty to seventy years. Now life expectancy is predicted to rise to over ninety by the end of the century – an incredible fifty percent increase in human lifespans.

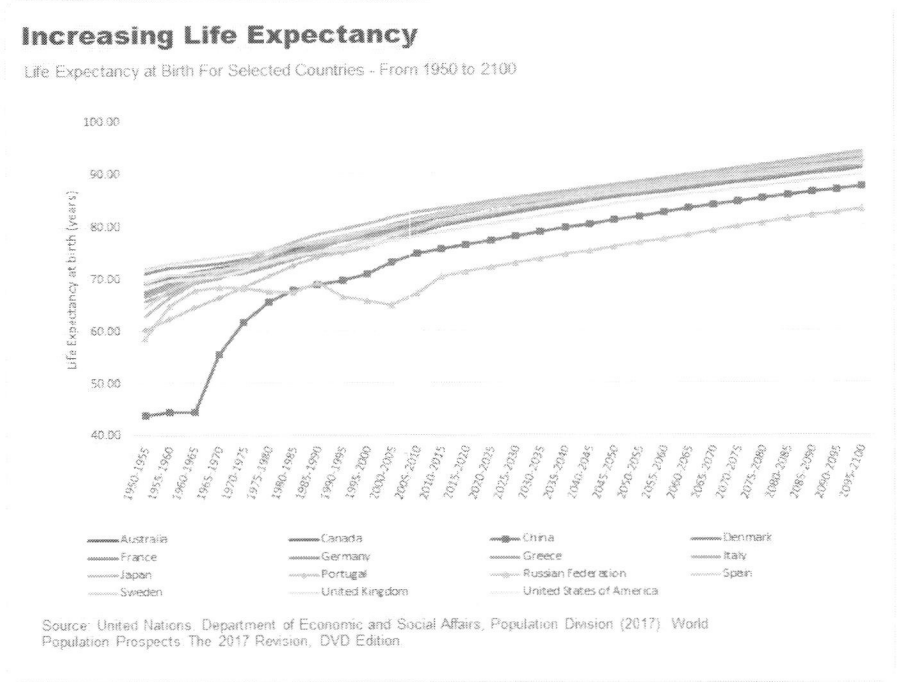

The abrupt plummet in fertility rates and the jump in life expectancy has pushed the world into a situation that is "without parallel in the history of humanity.[23]" according to the United Nations. This is because the world saw one of the largest population booms and a jump in life expectancy immediately followed by a dramatic and sustained population decline caused by falling fertility rates that can only be compared to the Black Plague in Europe or the fall of the Ming dynasty in China.

Looking at fertility rates in a broader view of history shows how unprecedented this really is. For centuries the world has seen sustained relatively high fertility rates –though some nations, like France during World War I or Germany during WWII, would briefly dip below the minimum population replacement rate. After the Second World War, fertility rates began to move closely in sync. First surging higher for the post-war baby boom, and then simultaneously crashing down beginning in the late 1960s.

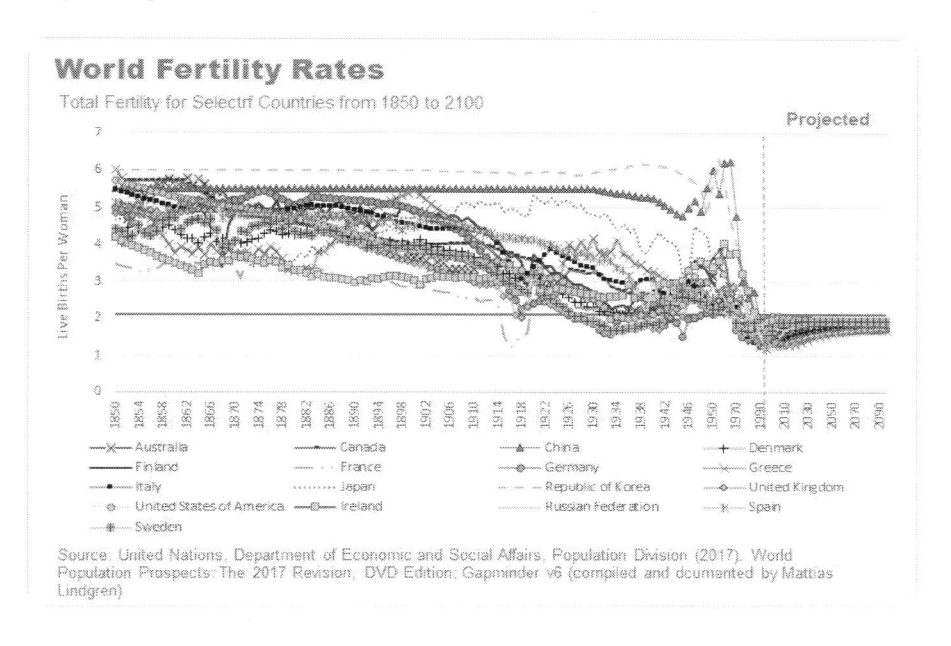

World Fertility Rates
Total Fertility for Select Countries from 1850 to 2100

Source: United Nations, Department of Economic and Social Affairs, Population Division (2017). World Population Prospects: The 2017 Revision, DVD Edition; Gapminder v6 (compiled and documented by Mattias Lindgren)

After centuries, if not millennia, of unabated growth, the collapse in fertility rates that began in the late 1960s will bring the trend of population growth to an abrupt end. Looking at a graph of the combined population of The United States, China, India, Italy, Japan, The UK, Germany, Russia* Spain, France, Mexico, Belgium, Australia, Brazil, South Korea, Poland, and Canada (who together represent 74.5% of the World GDP) shows the imminent and unparalleled population plummet in our future.

The End of Exponential Growth

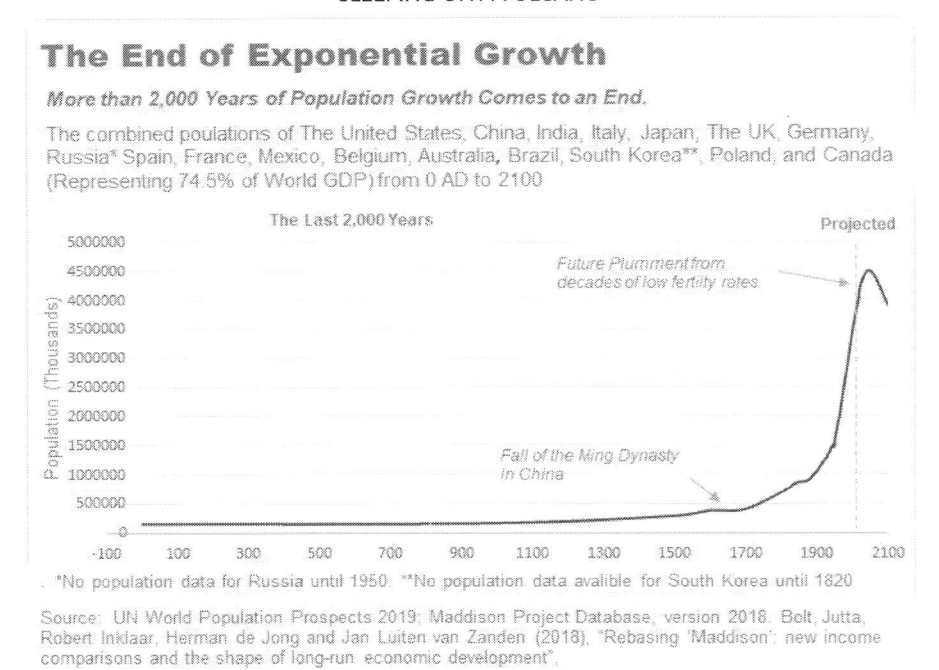

More than 2,000 Years of Population Growth Comes to an End.

The combined populations of The United States, China, India, Italy, Japan, The UK, Germany, Russia* Spain, France, Mexico, Belgium, Australia, Brazil, South Korea**, Poland, and Canada (Representing 74.5% of World GDP) from 0 AD to 2100

*No population data for Russia until 1950. **No population data avalible for South Korea until 1820

Source: UN World Population Prospects 2019; Maddison Project Database, version 2018. Bolt, Jutta, Robert Inklaar, Herman de Jong and Jan Luiten van Zanden (2018), "Rebasing 'Maddison': new income comparisons and the shape of long-run economic development",

While this graph of total population illustrates the magnitude and the unprecedented nature of the future decline in population, it is not the fall in total population that will have the first economic impact – before the decline in the total population the crucial working-age population will fall first.

A graph of the working age population of Southern Europe – the region that includes nations such as Italy, Spain, Portugal, Greece, and Croatia – illustrates the impact of the fall in fertility rates and the scale of the demographic shift facing the world.

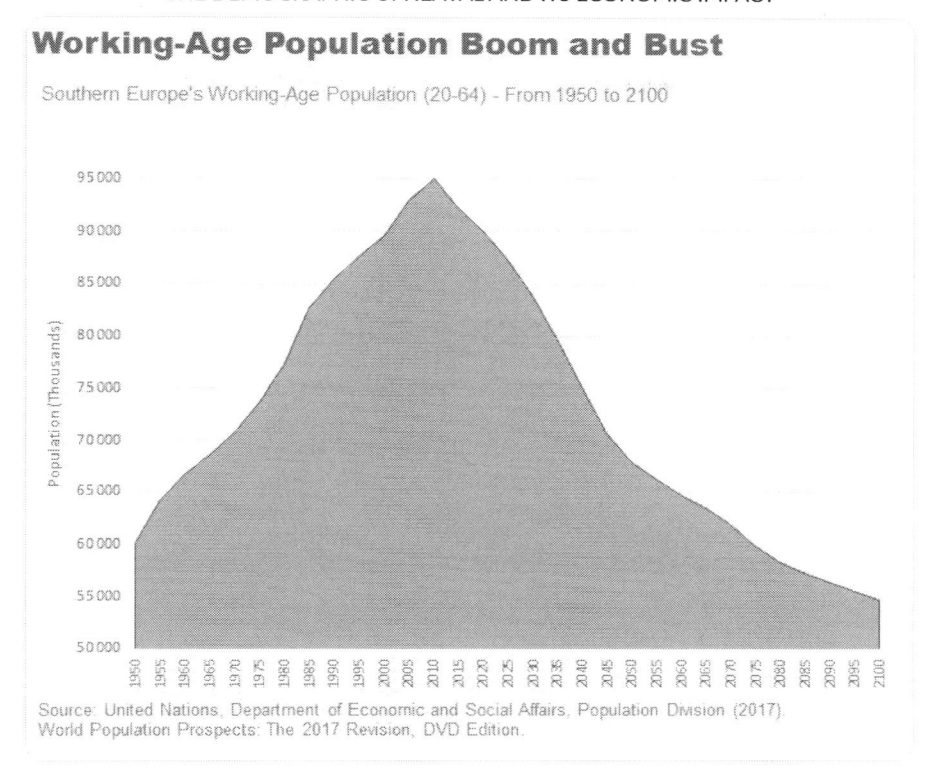

Working-Age Population Boom and Bust

Southern Europe's Working-Age Population (20-64) - From 1950 to 2100

Source: United Nations, Department of Economic and Social Affairs, Population Division (2017). World Population Prospects: The 2017 Revision, DVD Edition.

The population of Southern Europe between the ages of 20-64. Note the almost doubling of this age group from 1950 to 2010. After 2010, this group drops sharply for the foreseeable future.

And although the number of working-age people has already begun to fall sharply, increasing lifespans and the size of the *Baby Boomer* generation means that the number of older people will continue to increase until 2050. This lag will continually narrow the dependency ratio (the ratio between working-age people and the young and old which depend on them) of the developed world. Particularly as while *Boomers* are living longer they are not doing so in good health. The Boomer generation has significant increases in diabetes, obesity, and disabilities in general over previous generations.[24]

Southern Europe's Demographic Shift

Southern Europe's Population by Age Group - From 1950 to 2100

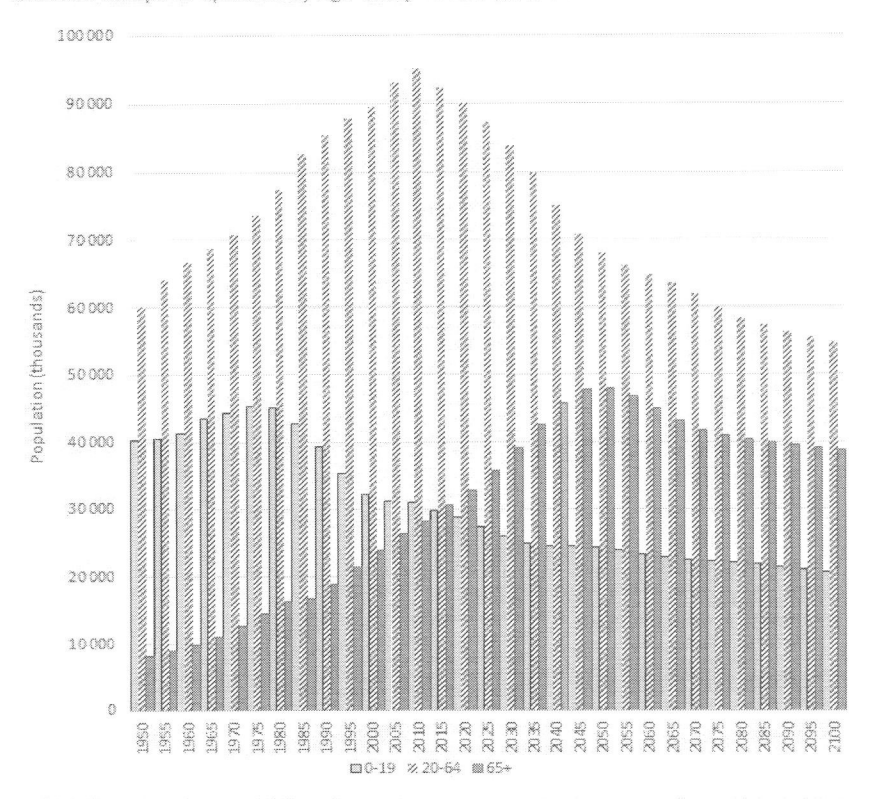

[Bar chart showing population in thousands by age groups 0-19, 20-64, and 65+ from 1950 to 2100]

Source: United Nations, Department of Economic and Social Affairs, Population Division (2017). World Population Prospects: The 2017 Revision, DVD Edition.

The *Baby Boom* was a worldwide phenomenon, as was the following plummet in fertility rates. But not all countries experienced the fall in fertility rates as sharply as others. The United States is one of the few major developed countries whose average fertility rate did not fall too far below the population replacement rate. The *Millennial* generation in the United States is roughly the same size as the *Boomer* generation – slightly larger even with immigration. However, the United States will still see a dramatic increase in the ratio between young and old as the massive Boomer generation retires.

For many countries though, they face not merely a rise in the ratio of retired to working-age people. They face an absolute collapse in their populations. Europe and East Asia in particular have had fertility rates far below the minimum replenishment rate for generations. They face a future of depopulation.

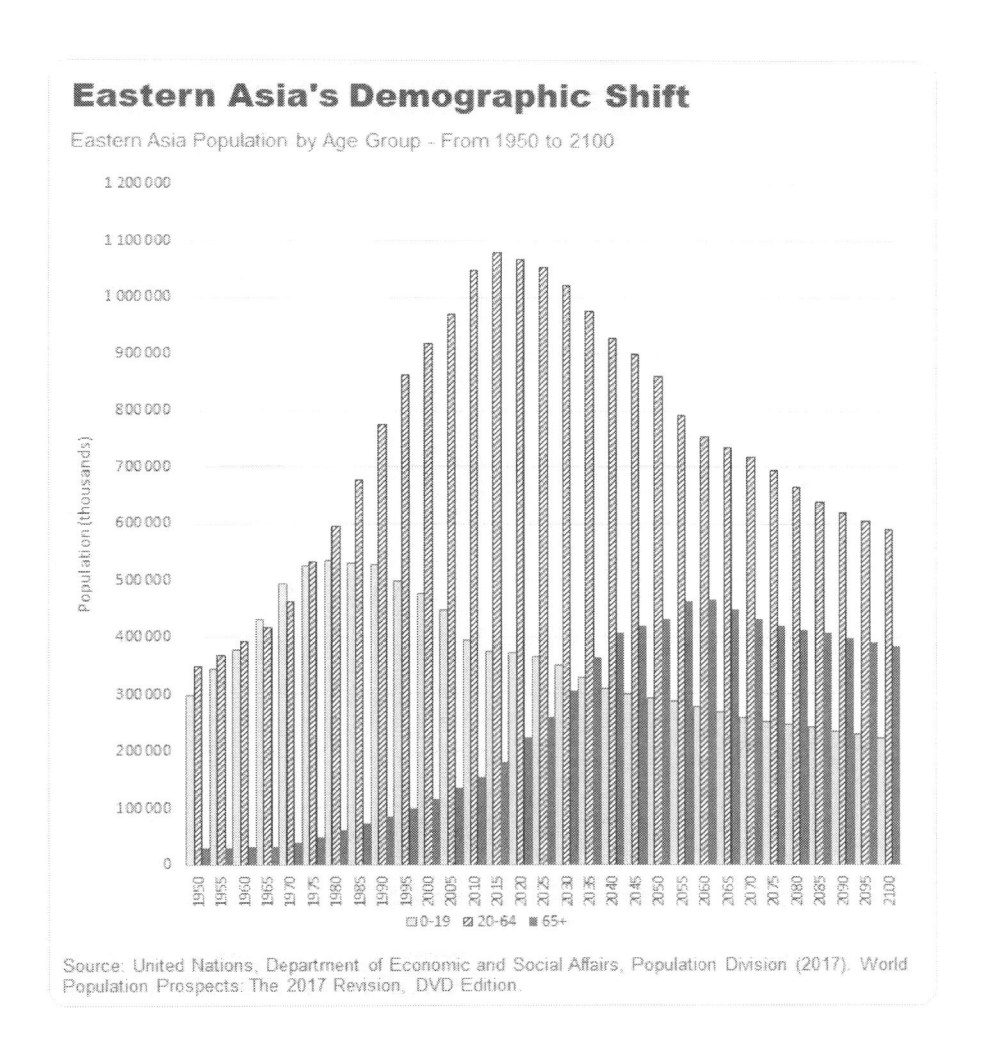

Eastern Asia's Demographic Shift

Eastern Asia Population by Age Group - From 1950 to 2100

Source: United Nations, Department of Economic and Social Affairs, Population Division (2017). World Population Prospects: The 2017 Revision, DVD Edition.

Population by age group in East Asia. Note the continuing narrowing of the gap between the working-age group and the elderly group as the working age population falls.

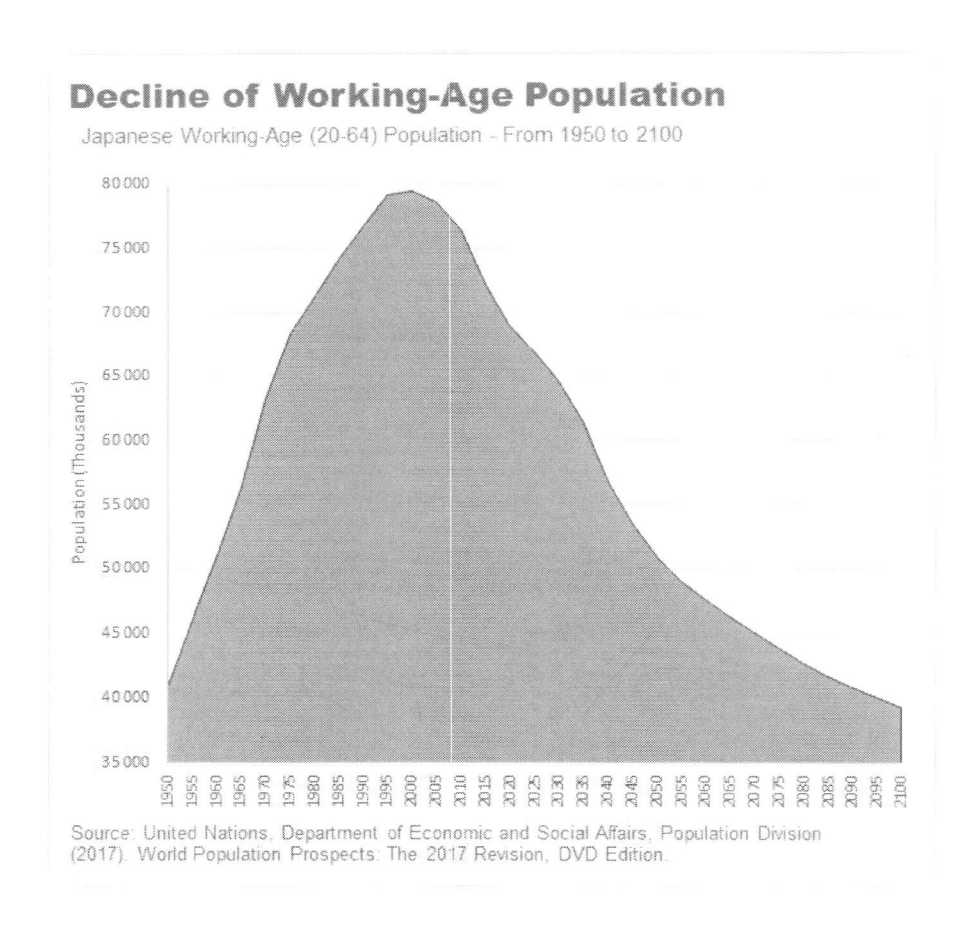

Decline of Working-Age Population
Japanese Working-Age (20-64) Population - From 1950 to 2100

Source: United Nations, Department of Economic and Social Affairs, Population Division (2017). World Population Prospects: The 2017 Revision, DVD Edition.

While the working-age population will return to 1950 levels in many countries, the elderly population will be much different. Compare for example, the population pyramids of Japan and Italy in 1950, and what their populations will look like in 2050.

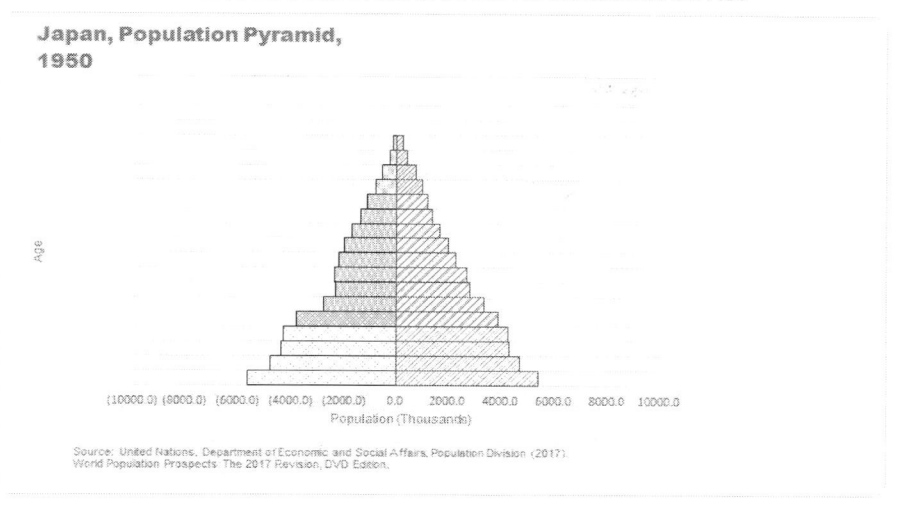

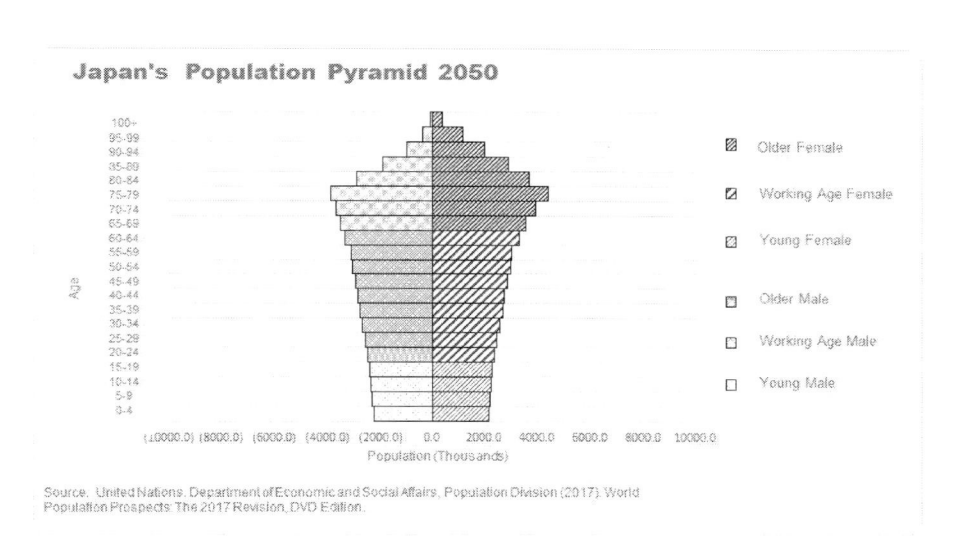

THE DEMOGRAPHIC UPHEAVAL AND ITS ECONOMIC IMPACT

Japan, Population Pyramid, 1950

Source: United Nations, Department of Economic and Social Affairs, Population Division (2017). World Population Prospects: The 2017 Revision, DVD Edition.

Japan's Population Pyramid 2050

- Older Female
- Working Age Female
- Young Female
- Older Male
- Working Age Male
- Young Male

Source: United Nations, Department of Economic and Social Affairs, Population Division (2017). World Population Prospects: The 2017 Revision, DVD Edition.

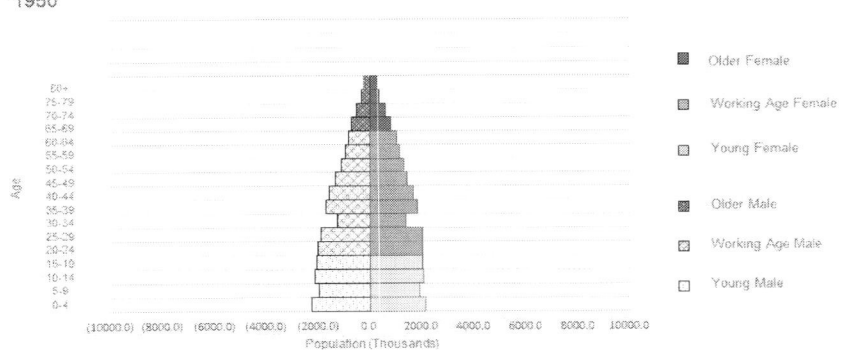

SLEEPING ON A VOLCANO

Italy, Population Pyramid, 1950

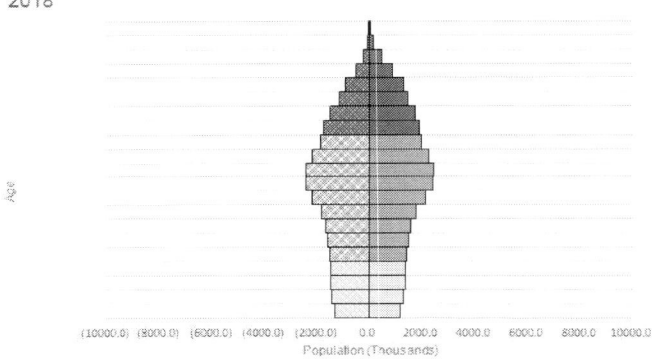

Legend:
- Older Female
- Working Age Female
- Young Female
- Older Male
- Working Age Male
- Young Male

Source: United Nations, Department of Economic and Social Affairs, Population Division (2017). World Population Prospects: The 2017 Revision, DVD Edition.

Italy, Population Pyramid, 2018

Source: United Nations, Department of Economic and Social Affairs, Population Division (2017). World Population Prospects: The 2017 Revision, DVD Edition.

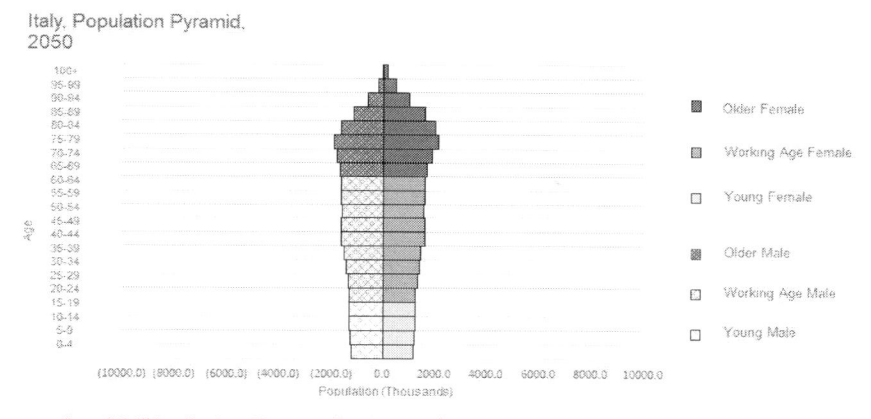

THE DEMOGRAPHIC UPHEAVAL AND ITS ECONOMIC IMPACT

Italy, Population Pyramid, 2050

Source: United Nations, Department of Economic and Social Affairs, Population Division (2017). World Population Prospects: The 2017 Revision, DVD Edition.

The United State's fertility rate fell along with the rest of the world, but as mentioned before, it hovered around 2.1 and therefore the United State's population pyramid will not invert like so many other nations. However, centuries of each generation being exponentially larger than the one before has come to an end. While the Baby Boomers supported a generation much smaller than themselves, the generations that followed the Boomers were approximately the same size and so the dependency ratio in the United States will rise sharply – though not as much as in other nations.

SLEEPING ON A VOLCANO

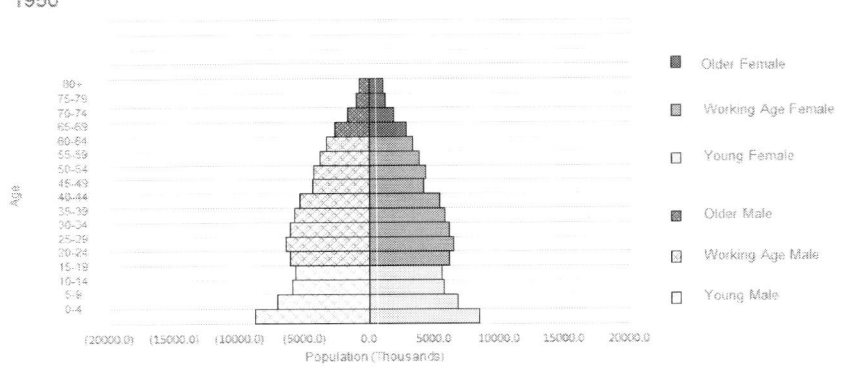

United Sates, Population Pyramid, 1950

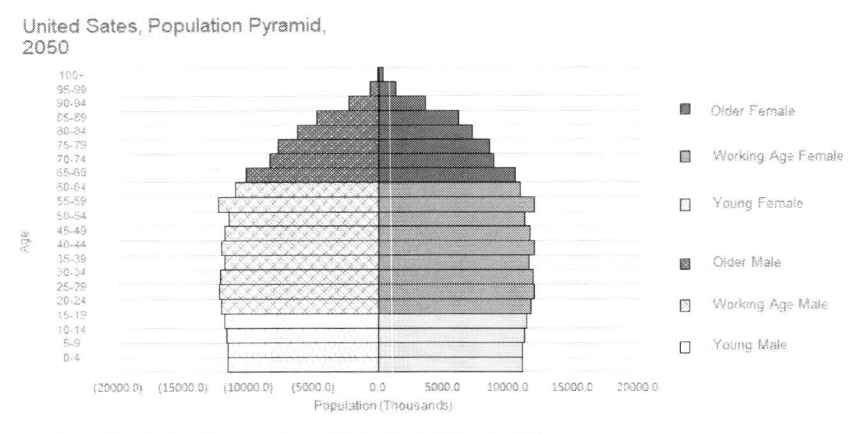

United Sates, Population Pyramid, 2050

Looking at the dependency ratio of various countries illustrates how the world has just entered the beginning stage of rapid aging. And this is not limited to a few select nations. Essentially every major economy is projected to age rapidly over the next few decades. With this aging, the ratio between working-age adults and the elderly over sixty-five will increase sharply.

Old Age Dependancy Ratio

The Number of Elderly / by the Number of Working-Age Adults For Selected Nations From 1950 to 2050

Number of Working-Age Adults for Every Person Over 65 in 1950 vs 2050

From 10 to 1.3
From 7.6 to 1.3
From 15.8 to 1.4
From 7 to 1.5
From 6.2 to 1.7
From 5.6 to 2.1
From 11.8 to 2.1*
From 7.0 to 2.5
From 11.5 to 2.5

- - - - United States
———— China
▬▬▬ United Kingdom
·········· Japan
———— South Korea
— ·· — Portugal
— — — Italy
— · — Russia
— ·· — Germany

Source: United Nations, Department of Economic and Social Affairs, Population Division (2017). World Population Prospects: The 2017 Revision, DVD Edition.

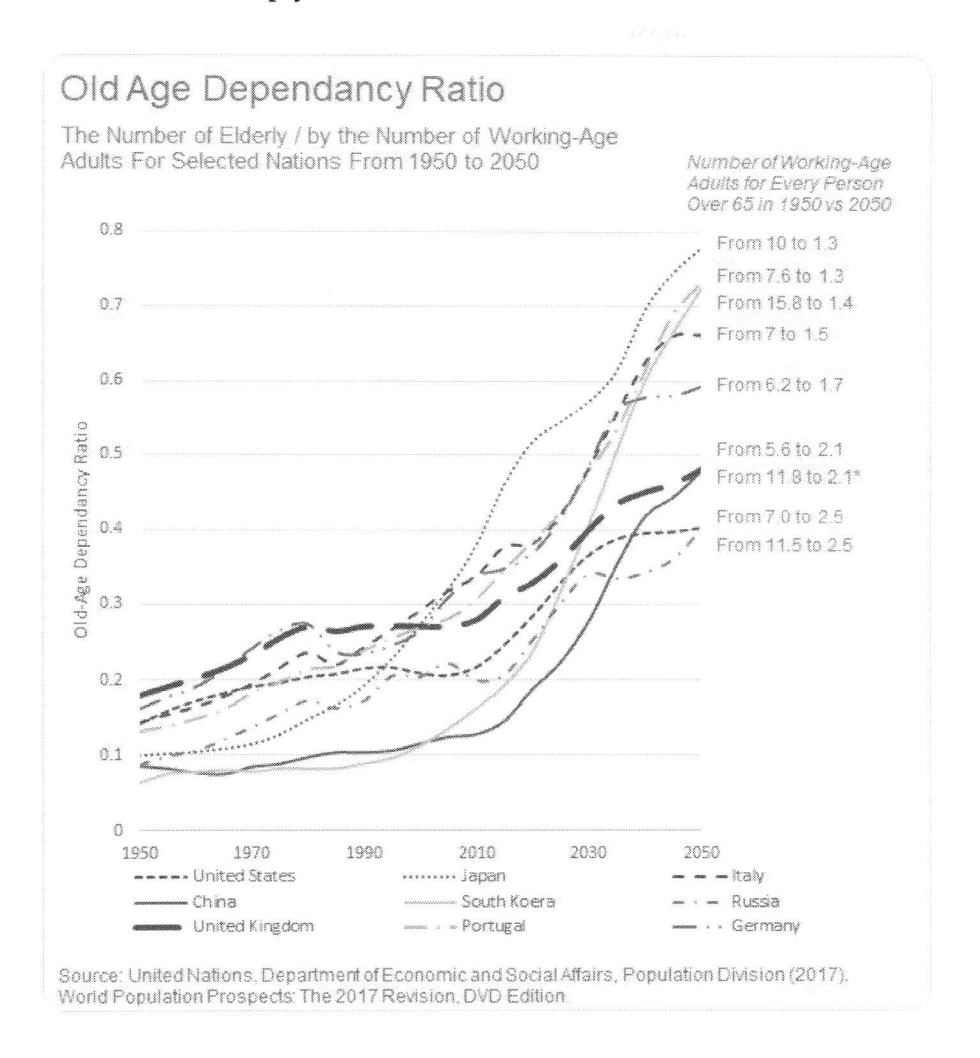

Take Japan for example – the third largest economy in the world – in 1950, about nine percent of their adults were over sixty-five. By 2050, almost forty-five percent of adults in Japan will be sixty-five or older. China, like much of the world, has spent almost seventy years with the percentage of their adults over sixty-five at about seven percent of the adult population. Over the next 20 years that number will rise to around thirty-two percent.

And these projections are certain. The generations have already been born – the die has been cast for decades. As former Secretary of Commerce and chairman of the *Council On Foreign Relations* Peter G. Peterson wrote back in 1999:

> *Over the next several decades, countries in the developed world will experience an unprecedented growth in the number of their elderly and an unprecedented decline in the number of their youth. The timing and magnitude of this demographic transformation have already been determined.* **Next century's elderly have already been born and can be counted — and their cost to retirement benefit systems can be projected.**
>
> **...there can be little debate over whether or when global aging will manifest itself.** *And unlike with other challenges, even the struggle to preserve and strengthen unsteady new democracies, the costs of global aging will be far beyond the means of even the world's wealthiest nations — unless retirement benefit systems are radically reformed. Failure to do so, to prepare early and boldly enough, will spark economic crises that will dwarf the recent meltdowns in Asia and Russia*[25].*[emphasis added]*

The Optimistic Nature of the Population Data

Throughout this book, the population data presented in based on the UN's 2017 median projected population data - what the United Nations population division believes is the most likely scenario. They also publish a low-variant and a high-variant projection which account for the uncertainty in their estimates. As pointed out earlier, most of the generations we are concerned with have already been born there is not uncertainty there – the projections are only for future unborn generations. However, there are excellent reason to believe that the UN median variant is overly optimistic. That future projections for birthrates to halt their fall and stabilize may be unlikely and that birthrates will continue to plunge. This argument is made in books such as *Empty Planet* which cities the work of demographers such as Wolfgang Lutz of the IIASA (International Institute for Applied Systems Analysis) and others who are skeptical of the UN projections and believe that birthrates may fall quicker than expected. In fact, in June 2019 the UN revised it's projected birthrates down significantly.

Additionally, there is excellent reason to believe that the population data out of China specifically – like so much of the data out of China – is unreliable and inflated. Yi Fuxian, a senior scientist at the University of Wisconsin-Madison and author of *Big Country with an Empty Nest*, recently wrote in the South China Morning Post:

> *China's total fertility rate, or the number of kids per woman throughout her life, dropped below the watershed level of 2.1 in 1991, from which moment the population size of the next generation would be smaller than the current one, and the average total fertility rate was 1.36 in 1994–2018, according to data from census and surveys. However, the family planning authority in charge of the country's population control refused to believe the numbers and "adjusted" the rate to 1.6–1.8 and, accordingly, the official population size.*
>
> *For instance, the real total fertility rate in 2000 was 1.22, according to a census result, but the government revised it to 1.8. Accordingly, the country had 14.1 million new births in 2000, but the government revised the figure by 26 per cent to 17.7 million. A census, which is conducted every 10 years, should provide the truest picture of China's demographic situation. But for the 2000*

census, the government was unhappy about the original finding of 1.24 billion and revised it up to 1.27 billion.

If the UN projections are optimistic, and the Chinese population is smaller than believed, the road ahead will only be more difficult and the decline more dramatic.

Economic Impact of the Demographic Decline

For decades the world has been warned that the aging of the world would cause financial problems – yet the world no longer seems concerned. Ironically, as the aging of the world now begins in earnest, polling shows the concern for such issues has fallen to new lows. But the predictions were not wrong. And no steps were made to alleviate the issues – its only that the Baby Boomers have not yet retired. The process has in fact just begun. In the United States, the first of the Baby Boomers turned 65 in 2011. But the aging of the world will be a process that takes decades – and with each and every year for the foreseeable future the ratio between the working age-population and the retired Baby Boomers will shrink.

This massive lopsidedness between the size of the Boomer generation and those that followed has already had a major impact on the world economy – especially when combined with increasing globalization. Wages have stagnated in large part because of the massive labor supply shock that was the baby boom, the integration of Asia into the global economy, and with the fall of the USSR, the integration of Eastern Europe. In simple economic terms, the price for labor is a function of supply and demand – and the last fifty years has seen a massive increase in the supply of labor. Economists from the Bank of International Settlements (BIS), a bank for central banks, Charles Goodhart and Manoj Pradhan, writing in their seminal *Demographics Will Reverse Three Multi-Decade Global Trends,* explain that:

Between the 1980s and the 2000s, the largest ever positive labour supply shock occurred, resulting from demographic trends and from the inclusion of China and eastern Europe into the World Trade Organization. This led to a shift in manufacturing to Asia, especially China; a stagnation in real wages; a collapse in the power of private sector trade unions; increasing inequality within countries, but less inequality between countries; deflationary pressures; and falling interest rates[26].

In addition to the increase in the total labor supply, the changing age structure of the world itself was an important driver of growth through the *demographic dividend*. The phrase *demographic dividend* is economist jargon for the economic benefits that are inherent with falling fertility rates. The United Nations Population Fund (UNFPA) explains that the *demographic dividend* is "a boost in economic productivity that occurs when there are growing numbers of people in the workforce relative to the number of dependents.[27]" And this boost is not insignificant. Indeed, much of the rise of Asia can be explained by the historically unprecedented fall in fertility rates. Professor Andrew Mason, writing in *Demographic Transition and Demographic Dividends in Developed and Developing Countries*, explains that:

> *The detailed case studies of the miracle economies of Eastern and South-Eastern Asia provide compelling and consistent evidence that the demographic dividend was an important contributor to that region's economic success... Bloom and Williamson (1998) use econometric analysis to conclude that about one-third of Eastern and South-Eastern Asia's increase in per capita income was due to the demographic dividend*[28].

More recent studies – such as a 2008's *Demographic Change and Economic Growth in Asia* find as much as 40% of the East Asian economic miracles can be explained by the demographic dividend[29].

End of the Demographic Dividend

The fall in fertility rates and subsequent population decline is not inherently a reason for concern. In fact many, like Paul Ehrlich, are no doubt enthused. But in many ways the growth and prosperity the world has seen over the last fifty years is because of the decline in fertility rates. Think about how much of growth in the last half century has been from Japan and China and the other Asian *economic miracles* – and consider that as much as 40% of that growth was directly attributable to the increase in the ratio of the working-age population because of the size of the Boomer generation and the falling fertility rates which followed it. The trouble arises from the fact that the current design of national budgets require ever expanding growth. Because of this, the final economic repercussions of the falling fertility rates – once the working age population begins to fall – may be as extreme as the population decline itself.

For decades there have been warnings of the fiscal problems that will arise as the *Baby Boomer* generation retires[30]. But there is more to the issue than increasing pension and healthcare costs. Economists use *The Life-Cycle Hypothesis* to explain how the average person's saving and consumption vary with age. People in their twenties to thirties borrow to buy their first car or house. While forty to sixty-five-year olds save more of their income for retirement. Lastly, older people will stop earning and saving and live off of the capital they have accumulated.

Savings and Conpumption by Age

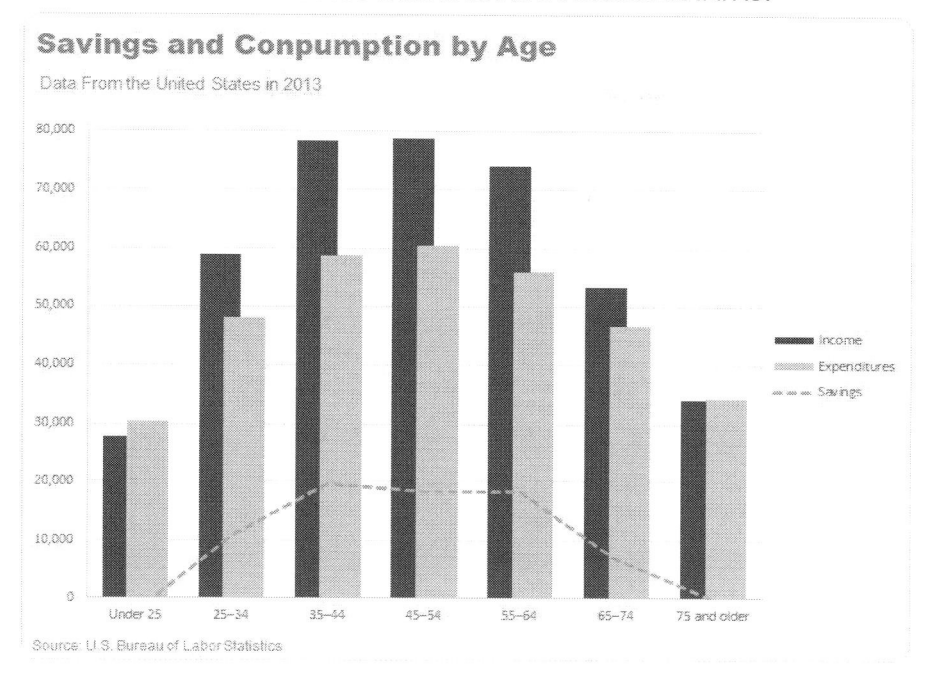

Data From the United States in 2013

Source: U.S. Bureau of Labor Statistics

This graph above captures some of the changes in consumption with age – but it is actually fairly misleading. It appears to show that older and younger people's income and expenses still roughly match. But this is only because the *income* of the young in this graph is a transfer from parents to children (or from student loans) and the income of the elderly is largely pensions (ultimately from working-age taxpayers) and savings. Looking at consumption from labor income and consumption by age shows the real net productivity of a nation is from people in their prime working ages between thirty and sixty-five.

Average Labor Income and Consumption by Age

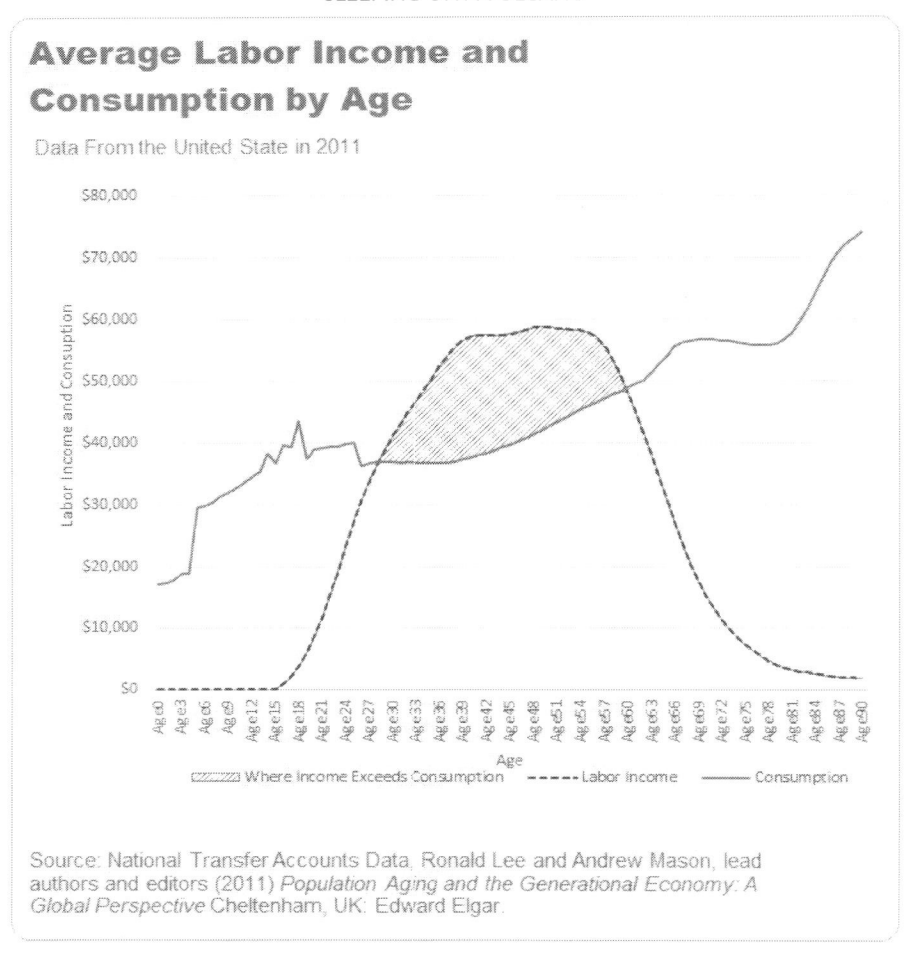

Data From the United State in 2011

[Chart showing Labor Income and Consumption by Age, with y-axis from $0 to $80,000 and x-axis from Age 0 to Age 90. Shaded area indicates "Where Income Exceeds Consumption". Dashed line represents Labor Income; solid line represents Consumption.]

Source: National Transfer Accounts Data, Ronald Lee and Andrew Mason, lead authors and editors (2011) *Population Aging and the Generational Economy: A Global Perspective* Cheltenham, UK: Edward Elgar.

At the moment the large Baby Boomer generation is still in their peak earning years and saving large amounts for their imminent retirement. Economist Jinill Kim of the Federal Reserve argues the retirement of the *Baby Boomers* will reduce the total amount of capital in the economy and thus slow GDP growth. He explains how:

> *... people move from being net borrowers in their youth to being net savers in their working years and finally to dis-savers*

[spending down their savings] in their elderly years. Therefore, if the share of elderly in the population rises, aggregate savings would fall, leading to lower investment growth, and, in turn, lower GDP growth...In most countries, the effects of demography on GDP growth rates are broadly negative in the projection period... the effect of the demographic factors is to make GDP levels peak in 2010 and then turn much lower by 2050, the end of the projection period[31]

Essentially, as the share of older people increases, the amount of savings and investment will fall. Rather than working and producing while saving and investing for retirement, the elderly stop producing and instead of investing they begin decreasing the total amount of investment in the economy by living off of their savings.

Rising Interest Rates

More than just slowing growth, the falling savings will also lead to higher interest rates. This is a more controversial position among economists, but there are excellent reasons to believe that the aging of the world will tend to *increase* real interest rates. While some economists have suggested that falling growth and investment opportunities would serve to depress interest rates, economists writing for BIS note:

> *Ageing will lower both desired savings and desired investment, but desired savings will fall by more. The resulting imbalance will require the real interest rate to rise for the market to clear. Just as the real interest rate has fallen since the 1980s thanks to a decline in desired investment borne out of the demographic sweet spot we described above, [demographic sweet spot being a creation of the baby boom followed by falling fertility rates] real interest rates will reverse course along with demographic trends and the*

resulting changes in savings and investment dynamics[32].

As the BIS economists note, interest rates are set by the supply of savings and the demand for investment. Rates have been falling for decades, adjusted for inflation, in large part because the large Baby Boomer generation has been saving more and more for retirement. But as they begin to retire, this will reverse. They will stop contributing to their 401(k)'s, stop investing in bonds and Treasuries for retirement, and begin slowly selling them off. This will reduce the total supply of savings in the economy – and thus interest rates will rise.

Fall In Demand for Bonds and Other Assets

The rise in interest rates is closely tied to another effect of the dramatic shift in the age structure of the world, a fall in demand for U.S. Treasury Bonds, and other assets.

Much of the demand for Sovereign debt over the last fifty years has been a consequence of the Baby Boomer generation saving for retirement. Private pension funds, state pension funds, mutual funds, and Social Security are all major holders of US government debt. Social Security is 100% invested in U.S. Treasuries.

We have seen how dis-proportionally large the Baby Boomer generation is. And we have seen how the life-cycle hypothesis dictate that people transition from being high earning tax payers to pensioners who live off their savings. What happens to the value of assets in the market when this generation stops being net purchasers of assets to net sellers?

One of the first serious analyses of the impact of the age shift and the Baby Boomers transition from being net savers and investors to living off their accumulated savings is Sylvester J. Schieber and John B.

Shoven's seminal 1994 paper *The Consequences of Population Aging on Private Pension Fund Savings and Asset Markets*. In the paper he concluded that:

> ... *private pensions will gradually cease being the major engine of aggregate saving that they have been for the past twenty years or more... By 2024, the pension system is projected to cease being a net source of saving for the economy. In fact, the pension system will then become increasingly a net dissaver. By 2040, the net real dissaving is more than 1.5 percent of payroll and by 2065 the negative saving is projected to reach almost 4.0 percent of payroll. This change of the pension system from a large net producer of saving to a large absorber of saving or loanable funds will likely have profound implications for interest rates, asset prices and the growth rate of the economy*

> *It should be emphasized that the timing of the prediction of the change in pensions from a net buyer of assets to a net seller is very sensitive to our assumptions about the rates of return earned on pension investments as well as to the assumed level of pension contributions. However, we feel that the pattern of Figure 3 [private pension system transitioning to being dissavers] is almost inevitable; only the timing could be somewhat different than pictured. If investment returns exceed our fairly conservative assumptions, then the decline of the saving contribution of pensions will be delayed in time. Still, the demographic structure is such that the decline will by necessity occur*[33].

A more recent 2010 BIS paper *Ageing and asset prices* by Előd Takáts takes another look at the same question.

> *Economic theory suggests that ageing affects asset prices negatively. As ageing economies account for the overwhelming major-*

ity of global investable assets, the question arises naturally: What will happen to asset prices? ... The results suggest that ageing will lower real house prices compared to neutral demographics over the next forty years in all countries in the sample. The estimated ageing impact is relatively mild in the United States with around 80 basis points per annum headwinds. The drag is estimated to be much larger in most of continental Europe and in Japan.

Furthermore, the results are also relevant when thinking about government debt sustainability. Two ageing related effects on government fiscal positions are well-understood. First, ageing will increase government expenditures, especially pension and health care spending. Second, ageing will slow economic growth as labor supply growth slows and in many cases reverses. Both of these factors will exacerbate government debt sustainability challenges. This paper highlights a third negative effect. Lower asset prices imply that long run interest rates will face upward pressure in the future. These higher long term interest rates would make debt sustainability even more challenging[34].

Surge in Costs

With the ageing of the developed world, growth will slow and interest rates will rise, while at the same time government's costs will increase sharply. Virtually all pension systems, including Social Security, are *pay as you go* systems. Which means that social security and similar systems are not paid to pensioners from the savings they invested in the system over their lifetime. That money was paid out to those already retired. In a pay as you go system, the pensions for retired people are paid by people who are still working. Which means that while the Baby Boomers supported a much smaller generation than themselves, the generations that followed must support the equally large, and in many countries larger, generation. These relatively smaller generations must support the future increase in pensions and healthcare.

Looking at spending in nations around the world, it is clear that spending has largely outpaced revenue for decades, with defense spending on the Cold War and entitlement programs contributing the bulk of expenses. With the end of the Cold War, the growth of entitlement programs are now driving spending in the developed world. The U.S. federal government spending is illustrative of this trend. The graph below shows the expenditures, as a share of GDP for entitlement programs – essentially Social Security, Medicare, and Medicaid.

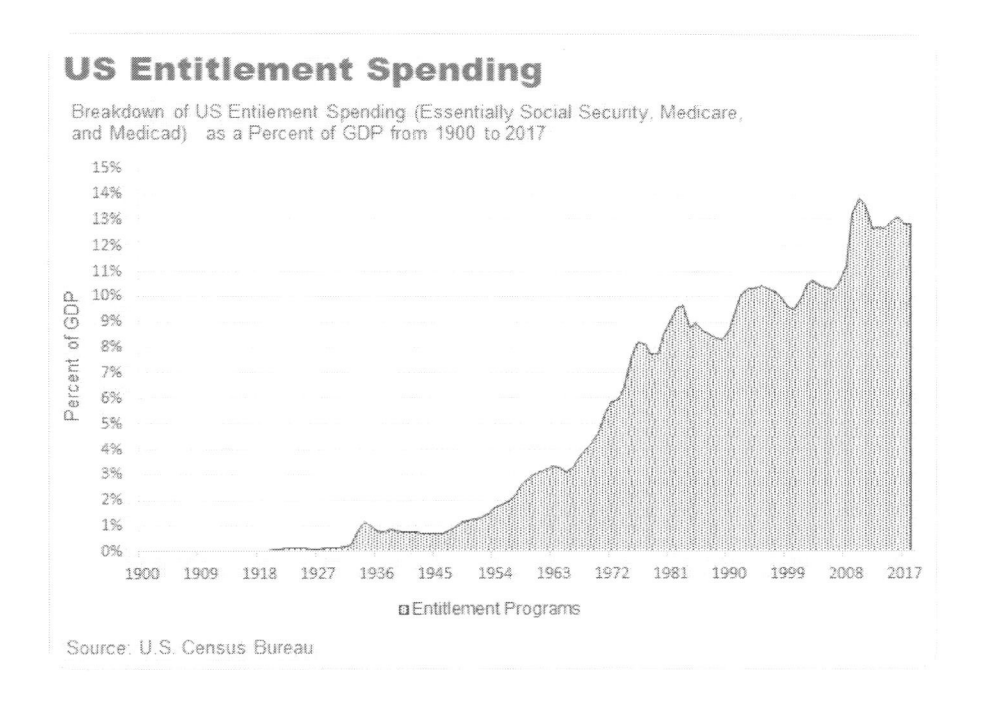

US Entitlement Spending

Breakdown of US Entitlement Spending (Essentially Social Security, Medicare, and Medicad) as a Percent of GDP from 1900 to 2017

Source: U.S. Census Bureau

Overlaying the chart with total revenues of the federal government, as a percent of GDP, shows the cause of the persistent deficits since the 1970s as entitlement rose dramatically as a percent of GDP.

US Spending and Revenue

Breakdown of US Entitlement Spending (Essentially Social Security, Medicare, and Medicad) and Revenue as a Percent of GDP from 1900 to 2017

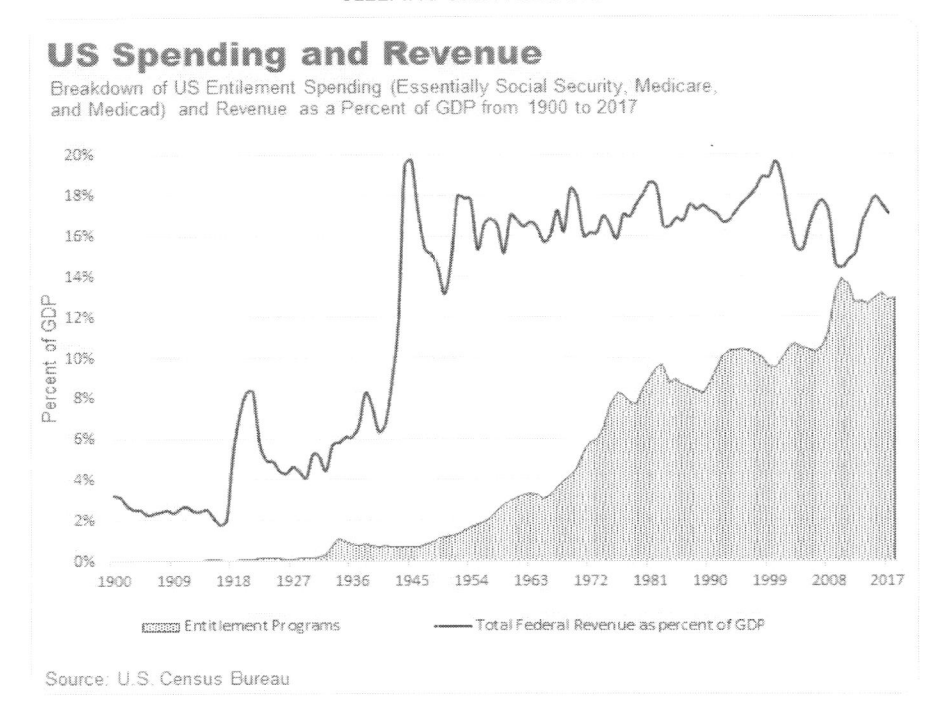

Source: U.S. Census Bureau

When the *Baby Boomer* generation retires, government obligations for pensions and healthcare will soar. In the United States, the large *Millennial* generation will at least stabilize the tax base. But in much of the developed world there was no analogously large *Millennial* generation. The tax base will shrink dramatically concurrent with the increase in mandatory spending in much of the developed world. Even in the United States, with relatively healthy demographics, mandatory entitlement spending (essentially social security, medicare and medicaid) and interest on the debt will soon exceed the total tax revenue of the federal government.

US Spending and Revenue

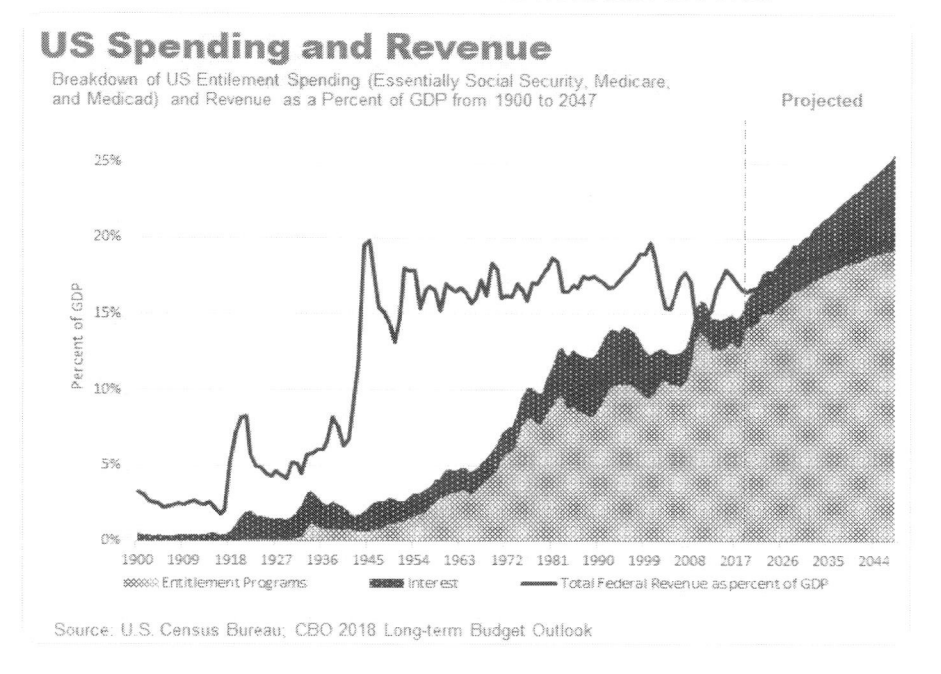

Breakdown of US Entitlement Spending (Essentially Social Security, Medicare, and Medicaid) and Revenue as a Percent of GDP from 1900 to 2047

Source: U.S. Census Bureau; CBO 2018 Long-term Budget Outlook

Spending on Entitlements from 1900 projected to 2048, alongside revenues projected to 2048. Note the sharp projected increase as the Baby Boomer generation retires and interest on the debt continues to grow. This projection, based on CBO data from early 2018, has been proved to be overly optimistic as entitlement spending slightly exceeds revenue now in early 2019.

Economic Shock

The economic risk to individuals from the population decline is not an inherent feature, but a result of the actions governments have taken. Nations have taken advantage of the large size of the Boomer generation to borrow enormous amounts. Nobel laureate Steven Chu recently laid out the issue very clearly, calling the world economy "a pyramid scheme." He explained that:

> *Increased economic prosperity and all economic models supported by governments and global competitors are based on having more young people, workers, than older people. Two schemes come to mind. One is the pyramid scheme. The other is the Ponzi scheme*[35].

Indeed, much of the world economy today has its roots in Baby Boomer generation being so much larger than the generations that followed it. Nations which decreased their fertility rates saw a substantial economic boost. Particularly in Asia where the fall was especially dramatic. Decades of low dependency ratios have been a driver of economic growth and allowed nations to create pay-as-you-go systems with generous pensions and healthcare. The savings of this generation also helped to push down real interest rates allowing countries to maintain higher debts.

But all of this will quickly reverse as this generation ages. Aggregate growth will slow - or even reverse. Costs for services will soar just as tax revenues plummet in many nations. And higher interest rates will slam the brakes on the debt accumulation of the last half-century.

3

The Rise and Fall of the Bretton Woods System

The size of the Baby Boomer generation and the falling fertility rates that followed it fundamentally changed the age structure of the world. This demographic shift has already had a significant economic and geopolitical impact; and the future reversal of population growth and the transition of the world from young to old will have an even larger impact. However, there is more to the crisis facing the world today than demographics. It is also necessary to understand the changes to the global economic system. The economic boost from the increase in population, and the demographic dividend from the low dependency ratios created by the fall in fertility rates, *allowed* nations to create pay-as-you-go systems and to build up massive amounts of debt. But the breakdown of the Bretton Woods monetary system in the 1970s created massive imbalances in global economy that incentizived, almost necessitated, the debt accumulation.

The global monetary system - how the currencies of the world function - is the bedrock of the capitalist system. Global trade, capital flows, inflation, interest rates, and other elemental parts of every economy in the world are impacted by, and often substantially determined by, monetary policy. Yet, many people do not understand how the international currencies function, or how the global monetary system has changed over the decades. To understand the looming sovereign debt crisis we need to understand the monetary system and how it has evolved.

The Bretton Woods Agreement

After the Second World War, only the United States remained as the military and economic superpower of the world. Every other country desperately needed loans or aid to rebuild their economy. This allowed America to dictate the terms of the post-war world. The idea of a *Rules Based International Order* refers to the institutions and norms that the United States created after 1945. The United Nations, The IMF, the World Bank, the Organization for Economic Cooperation and Development (OECD), NATO, the end of colonialism, and the birth of globalization and free trade – even the integration of Europe – were all pushed forward by the US in the aftermath of the world war. But the crown jewel of the post-war American initiatives was the Bretton Woods agreement.

To understand the Bretton Woods agreement and the need for it, you have to understand what happened to the monetary system after the First World War. During World War I the world went off of the gold standard in order to finance the war through inflation. And after the war, monetary instability increased with major inflation crises in France and Germany. The world attempted to return to the gold standard, but was never really able to do so. Even in the nations that did, countries sterilized their gold inflows[36] – breaking the spirit of the former gold standard[37]. Moreover, in the US, the Federal Reserve, under Benjamin Strong, kept interest rates artificially low after the war to reduce pressure on the Bank of England in their attempt to return to the gold standard at its pre-war exchange rate (a plan that led to deflation in England). These low rates helped keep gold in England, but also fueled a credit expansion in the United States. This expansion created a misallocation of resources and a tremendous economic boom – The Roaring Twenties[38].

When this bubble in the United States burst in 1929, the world fell into the deflationary spiral that was the Great Depression. In the previous century, the gold standard had worked to limit fluctuations in the money supply, but now it made it impossible to breakout of the deflationary spiral. Nations only halted the deflation once they again abandoned the gold standard and re-inflated their currencies.

The inflexibly of the gold standard, combined with the instability of the banking system, had proven to be a millstone around the neck of the world during the Great Depression. The abandonment of the gold standard by the United States and nations around the world ended the spiral of deflation. However, the gold standard had worked to automatically balance trade between nations. Without a peg to gold, exchange rates between currencies could be manipulated. Meaning that a nation could weaken their currency deliberately – lowering the value of their debts to other nations, and simultaneously boosting exports. Because of this, in the interwar period, tariff rates surged as nations attempted to protect their industries from the effects of manipulated currencies. This *Beggar thy neighbor* currency debasement or *competitive devaluation* of currencies led to trade wars and increased the economic turmoil and, perhaps worst of all, significantly damaged relations between nations.

In fact, the monetary instability that followed World War I in Germany – first hyperinflation and then deflation – is exactly what brought the Nazis to power. And the monetary changes that were instituted after the Nazi's came to power brought Germany out of the depression and solidified Hitler's control of the country.

The political fallout from the Great Depression was extreme in Germany - In the 1930 parliamentary elections, both German Communists and Nazi party made substantial gains, taking almost one-third of the popular vote between them. All parties became united in hostility to the major Western nations. The economic slump inflamed resentment against the allied powers and made it possible for the Nazi party to win the largest parliamentary share in the elections of 1932 - with Hitler appointed as Chancellor in January 1933.

The popularity of the Nazi party was solidified by the immediate recovery brought about by Hitler's finance minister - Hjalmar Schacht. Not a member of the Nazi party himself, Hjalmar Schacht was also responsible for ending the German hyperinflation almost a decade earlier. Liaquat Ahamed, writing in the *The Lords of Finance*, details how:

> *...immediately upon taking office, Schacht threw the whole baggage of orthodox economics overboard. He embarked on a massive program of public works financed by borrowing from the central bank and printing money*[39].

This stimulus broke the German economy out of its deflationary spiral - very similar to what Franklin Roosevelt accomplished in the United States. Unemployment immediately began falling. Over the next four years, Industrial production doubled and GNP rose dramatically.

The Recovery of the Germany Economy

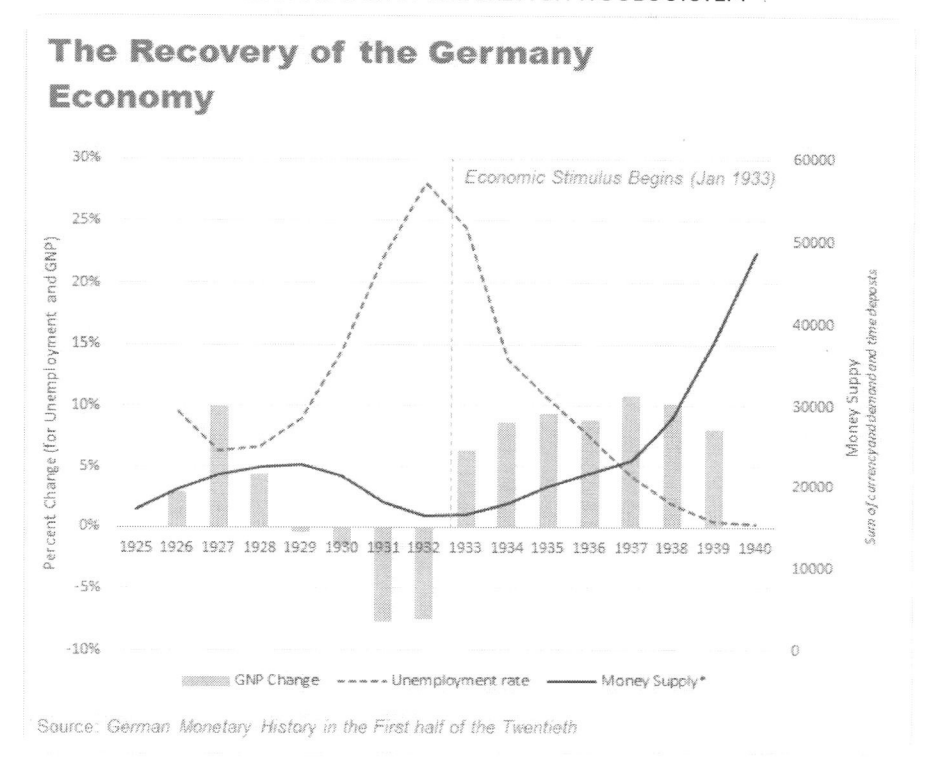

Figure: Chart showing GNP Change, Unemployment rate, and Money Supply for Germany from 1925-1940, with Economic Stimulus beginning Jan 1933.

Source: *German Monetary History in the First half of the Twentieth*

After the Second World War, the United States was determined that, unlike World War I, World War II would not be followed by economic turmoil which would risk sparking another major conflict. The United States believed that a system of free trade was vital to achieving both prosperity and peace in the post-war world. As President Roosevelt's Secretary of State, Cordell Hull, wrote in his memoirs:

> *I saw that you could not separate the idea of commerce from the idea of war and peace. You could not have serious war anywhere in the world and expect commerce to go on as before. And I saw that wars were often caused by economic rivalry. I thereupon came to believe that if we could increase commercial exchanges*

among nations over lowered trade and tariff barriers and remove international obstacles to trade, we would go a long way toward eliminating war itself[40].

Thus, the end of the Second World War brought not only peace, but also an enormous reduction in trade barriers – far below the levels seen even under the gold standard.

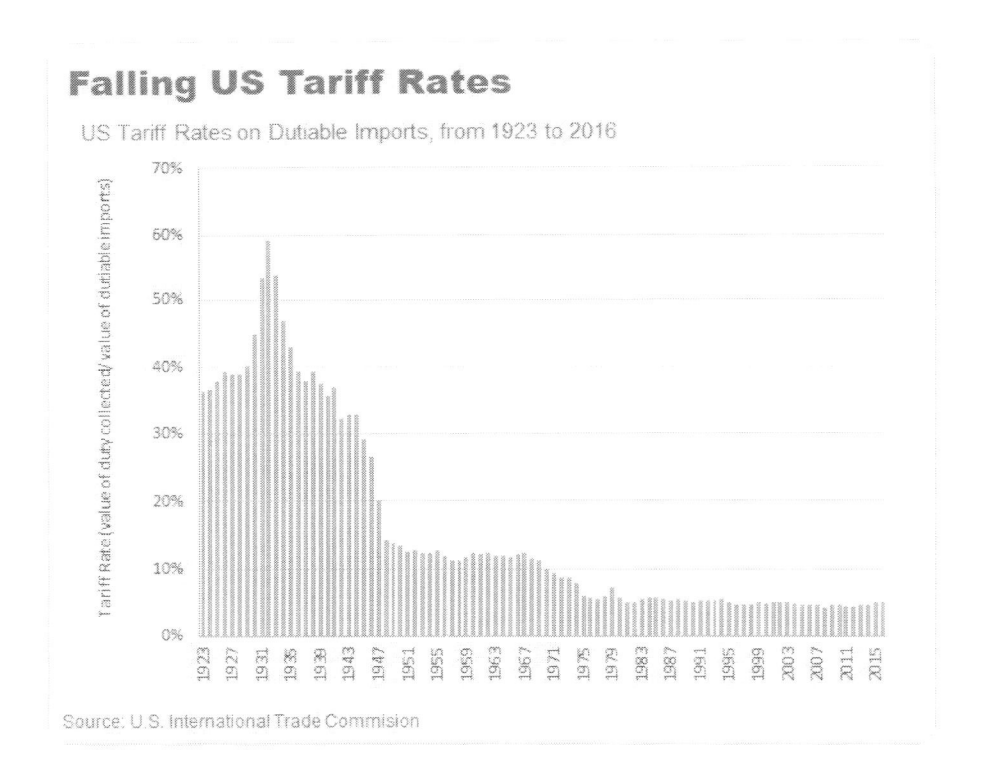

Falling US Tariff Rates

US Tariff Rates on Dutiable Imports, from 1923 to 2016

Source: U.S. International Trade Commission

But to make this dream of free trade economically feasible, the United States first addressed the root cause of the breakdown in trade before the Second World War –monetary instability and competitive currency devaluation.

Benn Steil, writing in *The Battle of Bretton Woods*, explains the position the United States found itself in at the end of the conflict:

> *The United States had been blessed by a unique confluence of events with a momentary window in which it could, in return for its now-vital financing service, not only put an end to competitive devaluation and trade protectionism...but permanently eliminate the old European powers as rivals and obstacles on the global stage*[41].

The United States used it's power at the end of the war to push forward a solution to the competitive currency devaluation that had sparked trade protectionism in the inter-war period-the Bretton Woods agreement. Bretton Woods served as a replacement for the gold standard and was the formal economic agreement for the antebellum. As economist Judy Shelton explained in her book *Money Meltdown*, under the Bretton Woods system:

> *The U.S. dollar was solidly anchored to gold, other countries currencies were pegged to the dollar at fixed exchange rates, and the resulting elimination of currency risk would serve as a tremendous boon to international trade and economic growth around the world*[42].

The Triffin Dilemma

While the world prospered under the Bretton Woods agreement, the system was unsustainable. As early as October 1959, Yale professor Robert Triffin informed the U.S. Congress' Joint Economic Committee that the Bretton Woods system was doomed. Triffin explained that the current monetary system could not survive because of the inherent imbalances it incentivized. The inherent unsustainablity was caused by the conflicting incentives which would become known as the *Triffin*

Dilemma.

Under Bretton Woods, the dollar was convertible to gold at $35 an ounce – but that did not mean that the United States had fixed the amount of currency it created. Instead, the United States could print currency as was needed to benefit the U.S. domestic economy, and, as the rest of the world was pegged to the dollar, other nations would be forced to expand their own currencies in step with the United States. This feature of the Bretton Woods system was famously characterized as an "exorbitant privilege" by French finance minister Valéry Giscard d'Estaing[43]. If this persisted, Triffin argued, eventually the supply of dollars would increase and the account balance of the country would deteriorate enough that the world would lose confidence in the dollar's peg to gold at $35 an ounce. Triffin argued that the country would be unable to hold the peg if confidence was lost and nations insisted on converting their dollars to gold. He argued that "A fundamental reform of the international monetary system has long been overdue. Its necessity and urgency are further highlighted today by the imminent threat to the once mighty U.S. dollar[44]."

Collapse

Largely ignored at that time, his warning became famous only when in 1971, as he had predicted, the Bretton Woods system collapsed. As Triffin had foreseen, The United States had printed a substantial amount of dollars since the end of World War II, far more than its gold reserves. In the late 1960s, having lost faith in the ability of the U.S. to keep the dollar-gold peg, Charles de Gaulle sent the French navy to exchange their dollars for gold, an act which was soon followed by other nations. By 1971, the ratio of dollars in foreign hands to gold in Fort Knox had declined precipitously.

On August 15th, 1971, President Richard Nixon addressed the nation in a prime time televised speech where he announced that the dollar would no longer be convertible to gold. Time Magazine later explained the decision was "... made to prevent a run on Fort Knox, which contained only a third of the gold bullion necessary to cover the amount of dollars in foreign hands". This de facto default marked the end of the Bretton Woods system. It also began to crash the value of the dollar - leading to a period of great inflation in the United States. Volcker took the federal funds rate up to an incredible 20% - eventually reducing the money supply and inflation expectations enough to curb the crashing value of the dollar.

The Imbalanced Global Economy

In the post-Bretton Woods era, all currencies which had been pegged to the dollar were now free-floating. Nations were free to increase the amount of their currency in circulation without regard to any pegs to the dollar. The dollar itself was also free-floating - severed from the peg to gold which it had been loosely holding.

While not immediately obvious, this had an incredible impact on the world - easily as significant as the unprecedented demographic upheaval - firstly through global trade. With the end of the Bretton Woods system there was no longer a mechanism to automatically balance imports and exports between nations as there had been under the gold standard and under the fixed exchange rates of the Bretton Woods system. This meant that countries could manipulate their currencies to boost exports, some nations created massive trade surpluses while other countries faced trade deficits and matching debt accumulation. Capital flows, production, trade, large portions of the global economy were shifted in ultimately unsustainable ways.

In addition to enabling massive trade imbalances, the free-floating system removed the external fiscal discipline of the gold standard or the fixed exchange rates of the Bretton Woods system. Under these systems, massive budget deficits would have an immediate negative impact which is not the case under the current free-floating system.

Global Trade

Not surprisingly, an abrupt and dramatic change in how international currencies functioned had a significant impact on world trade. As Brian Reinbold and Yi Wen of the Federal Reserve of St. Louis explain:

> *The long-running U.S. trade deficits and the emergence of China as a major creditor nation to the U.S. seem to be the result of two major economic forces: (1) the breakdown of the Bretton Woods system, which caused the U.S. currency and U.S. government debts to become the world currency and a global form of liquidity and store of value; and (2) the shifting of comparative advantage in goods production, which caused the reallocation of labor-intensive manufacturing from the U.S. to nations with cheaper labor*[45].

We will see that this "shifting of comparative advantage... to nations with cheaper labor" is also related to the breakdown of the Bretton Woodsy system.

Free-floating currencies have created massive trade imbalances which have been a major driving force of the past half century as these surplus nations have risen in power and wealth while the deficit countries have stagnated and accumulated debt. We saw that demographics played a large role, but the other part of the picture are these trade imbalances that incentivize debt accumulation. These imbalances can only be explained if we understand how trade functions under free-floating currencies.

Firstly, there has been an enormous amount of volatility in the currency exchanges under the system of free-floating exchange rates. Nobel prize winning economist Robert Mundell, In a 2012 paper urging currency reform, condemned the flexible exchange rates for its impact on trade. He wrote:

> *The flexible exchange rate experiment has been a failure. The best test of any monetary system is the degree to which it avoids unnecessary changes in real exchange rates.* ***These changes drastically reduce the gains from trade and disqualify the arguments ordinarily made for free trade areas and the customs unions****. By this criterion, the worst period in history has been the period since generalized floating that began in 1973. All of the arguments made for flexible exchange rates have proved to be incorrect. Destabilizing capital movements have rocked the exchange rates between areas that have a high and consistent degree of price stability. Exchange rates consistently overshoot equilibrium, causing harmful shifts between traded and non-traded goods industries and in the levels of indebtedness of rich and poor countries . [emphasis added]*[46]

But, free-floating currencies have been much more damaging to the economy than merely because of the increased volatility. Economists understand that trade is not a zero sum game, meaning that trade does not result in winners and losers - everyone can benefit from trade between nations. Greater trade allows for increased specialization and increased total global output. Classical economists, such as David Ricardo, proved centuries ago, that an increase in trade would lead to benefits for all. Even if one nation was better than another was at producing every kind of good, comparative advantage means that trade allows them to focus even more resources on what they are *best* at - where they have a comparative advantage. Thus, trade moves workers and resources into industries where they are most productive.

Therefore, while there might be some short-term disruption, in the long-term everyone is better off. However, while this classical analysis is indisputably logical, there is a hidden assumption in the analysis that is no longer valid in our modern world. The theoretical foundations of the classical argument for free trade depends on the trade and exchange taking place with currencies which are both pegged to gold (or some other fixed exchange rate).

Under the current free-floating monetary system, the prices of currencies themselves can be manipulated - which affects the calculation of comparative advantage for trade. Currencies can be undervalued or overvalued relative to each other. One might logically suppose that when six Turkish Lira can be exchanged for one American dollar, the actual purchasing power of one dollar would be roughly the same as six Turkish Lira. In other words, if a cheeseburger cost $5 in the United States, when you exchange currency you may get ¥500 yen, but a cheeseburger in Japan should then cost roughly ¥500 yen. But this is not the case. Many nations manipulate the currency exchanges to undervalue the purchasing power parity (PPP) of their own currencies. Making everything priced in their currency cheaper to buyers from foreign countries. This has had an enormous impact because, as the former Federal Reserve chairman Paul Volcker said, "A nation's exchange rate is the single most important price in its economy. It will influence the entire range of individual prices, imports and exports and even the level of economic activity."

Consider how trade works with free-floating currencies. One country can manipulate their currency lower so that everything they produce is significantly cheaper from the perspective of other countries. This distorts the idea of comparative advantage because prices are no longer reflective of the underlying economic reality. Prices are no longer a function of supply and demand or relative efficiency - with manipulated currencies it is impossible to even know if resources are being allocated more efficiently or not.

Think about it. Let's say one country actively manipulates their currency exchange rate so that all of a sudden everything in their country is cheaper to buy with dollars. Canada could do it tomorrow – the bank of Canada wants a cheaper Canadian Dollar so they go and buy US dollars and sell Canadian dollars. Did they suddenly gain a greater comparative advantage? Will they be able to produce more efficiently? Regardless, Canadian exports will rise and production will move to Canada regardless of if that is actually the most efficient allocation of resources.

Currency Manipulation Does Affect the Trade Balance – Through Indirectly.

Now, there actually a great deal on confusion around this issue. Some economists will tell you that currency exchange rates are inconsequential for trade. For example, the 1998 testimony to the Subcommittee on International Economic Policy and Trade Committee on International Relations, where Daniel Griswold of the CATO Institute said:

> *The most important economic truth to grasp about the U.S. trade deficit is that it has virtually nothing to do with trade policy. A nation's trade deficit is determined by the flow of investment funds into or out of the country. And those flows are determined by how much the people of a nation save and invest – two variables that are only marginally affected by trade policy* [47].

On the first point, he is correct. Trade flows are determined by how much the people of a nation save and invest. But, in actuality, such capital flows are highly impacted by trade policy.

To understand the link between the national savings rates and the trade balance, you need to look at the economy as a whole. In economist lingo the national savings S minus the total investment I is equal to the total exported goods X minus the total imports M. with the formula being: S−I = X−M. That may seem a bit confusing − but think of it this way, if a nation produces a certain amount of value − they can either consume that value, invest than value, or export than value. The less value a nation consumes, the higher their savings rate is, the more they have available for investment and exports.

So, Canada's trade balance − the balance between their imports and their exports − is really a function of their levels of savings and investment. Canada produces a certain amount of goods and services, but imagine that they have a fantastically high savings rate and they consume little of whay they produce, what happens to the goods and services that they created? That value must either be consumed, invested, or exported. So, the higher their savings rate is, the higher exports will tend to be. And of course the reverse is true − if Canada consumes more than it produces it must obviously import goods and services.

It is this analysis that causes some to argue that tariffs and currency manipulation will have no impact on the trade balance of a nation, because while an artificially low currency might change comparative advantages, it is the savings and investment rate that affects the trade balance on the whole. But what is missed in this analysis is that the high savings rates, as seen in China for example, are the result of policies that suppress domestic consumption − like an artificially low currency.

As Michael Petteis describes in his fantastic book, *The Great Rebalancing*, in which he explains in detail how nations have created policies which suppress consumption − thus boosting exports, ultimately creating the massive imbalances we see today between nations. He writes:

Many Asian countries have followed the growth model established in the 1960s and 1970s by Japan, and this growth model includes crucially these three conditions:

1. Systematically undervalued currencies, in which the central bank intervenes in the currency to reduce its exchange value

2. Relatively low wage growth, in which wages grow more slowly than improvements in worker productivity

3. Financial repression, in which the state allocates credit and the central bank forces interest rates to below their natural or equilibrium rates[48].

All of these actions suppress household consumption – moving value from the household sector to the goods producing sector. As he explains:

...an undervalued currency, by raising the cost of imports, acts as a kind of consumption tax for household and so reduces disposable household income. With lower disposable household income usually comes lower household consumption...the combination of lower consumption and higher production automatically causes a surge in the savings rate[49].

And of course, this also acts as a subsidy to consumption for other countries whose currencies can now purchase additional imports.

It is important to realize that on the global scale trade surpluses and trade deficits must be balanced. So every trade surplus in one nation causes a trade deficit in another nation. If Canada dramatically increases its savings rate tomorrow, all else being equal, this will cause them to export more and some other nation to import more.

The Asian Growth Model

With the consumption suppressing/exported driven growth models enabled by free-floating currencies, trade imbalances have become a defining feature of the last fifty years. Lord Mervyn King, former Governor of the Bank of England, described how today's massive trade imbalances have divided the world:

> *The surpluses are concentrated in four countries – the euro area, China, Japan and Korea. Taken together, their combined current account surplus in 2016 was $886 billion, just over 3% of their GDP. The deficits were also concentrated in four countries – the United States, United Kingdom, Canada and Australia. Their combined deficits were $680 billion, just under 3% of their GDP. It is a striking example of the difference between the two "groups of four": the Anglo-Saxon world, with its instinct of openness to trade and competitive financial markets, and Continental Europe and the Far East, with a more mercantilist outlook*[50].

Under the current free-floating monetary system, imports and exports are no longer balanced between nations. Trade is no longer balanced by gold (which despite its faults did balance trade) or through the fixed exchange rates of the Bretton Woods system that was created after World War II. Today a country can manipulate its currency lower and run trade surpluses for decades - leading to a larger and larger trade deficits.

This is precisely what has happened to the United States over the last fifty years since the end of Bretton Woods. A massive trade deficit was created in the United States. And a tremendous amount of U.S. manufacturing jobs were outsourced - with almost no counterbalancing increase in other sectors on the U.S. Economy. Such an outcome defies classical economic logic - but is possible because of the breakdown of the Bretton Woods monetary system in 1971.

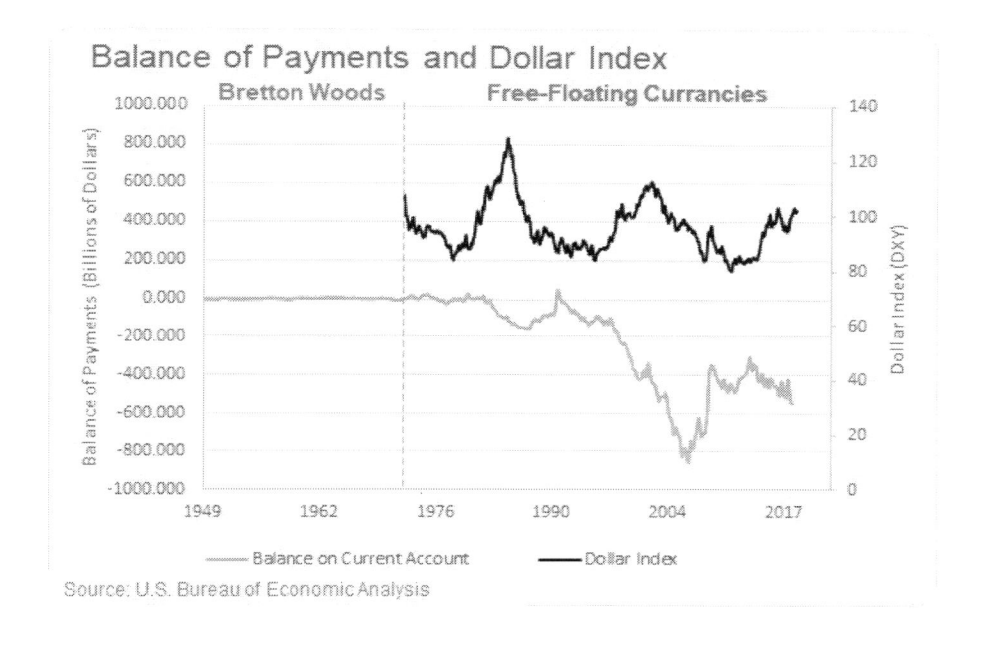

Balance of Payments and Dollar Index

Source: U.S. Bureau of Economic Analysis

United States Trade Deficit, 1960-2017

Sources: U.S. Census Bureau, Economic Indicator Division; Bureau of Economic Analysis

Free-Floating Currencies and Debt.

The emergence of free-floating currencies incentivized debt accumulation because (1) interest rates were lower as debt was seen insulated from the risk of default (2) There was no longer an incentive to balance budgets – every country could stimulate their economies with low interest rates and have the added benefit of boosting their exports as a relatively weaker currency would make their goods more attractive. (3) Debt rose significantly because the trade imbalances created by some nations engaging in t currency manipulation forced other nations to either engage in a currency war on take on debt.

As Mechael Pettis shows in his brilliant book, *The Great Rebalancing*, using Germany and Spain examples to explain how German trade surplus will cause Spanish debt to rise, he writes:

> *As long as Germany runs current account surpluses for many years and Spain the corresponding deficits, it is by definition true there mush have been net capital flows from Germany to Spain as German individuals and institution bought Spanish assets or lent them money to balance the current account imbalances. The Capital and current accounts for any country, and for the world as a whole, must balance to zero.*
>
> *...In the old days of specie currency – gold and silver –this meant that specie would have flowed from Spain to Germany as the counter balancing entry, and of course the flow created its on resolution. Less gold and silver in Spain relative to the size of its economy was deflationary in Spain and more gold and silver in Germany was inflationary there. Spanish prices would have declined and German prices risen to the point where the real exchange rate between the two countries would have adjusted sufficiently to reverse the trade imbalances...Large current account surpluses and deficits, in other words, could not persists because they were limited by the gold and silver holding of the deficit countries...In today's world things are different. There is no adjustment mechanism...that permits or prevents persistent current account imbalances*[51].

These forces meant that the end of the Bretton Woods system meant the end of balanced budgets and the beginning of a massive accumulation of debt around the world.

US Budget Deficit/Surplus in Constant 2018 Dollars, 1950-2018

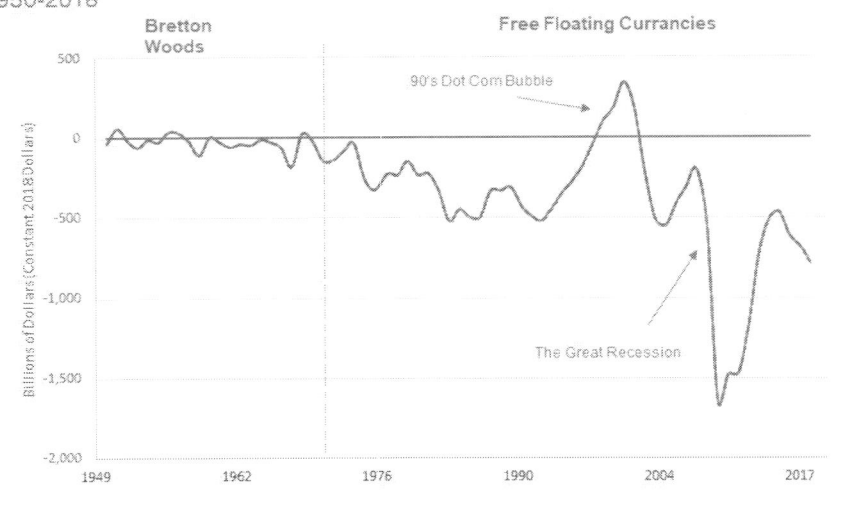

As budget deficits in the US increased the debt, which had been steady in total real terms and falling as a percentage of GDP began to increase substantially.

THE RISE AND FALL OF THE BRETTON WOODS SYSTEM

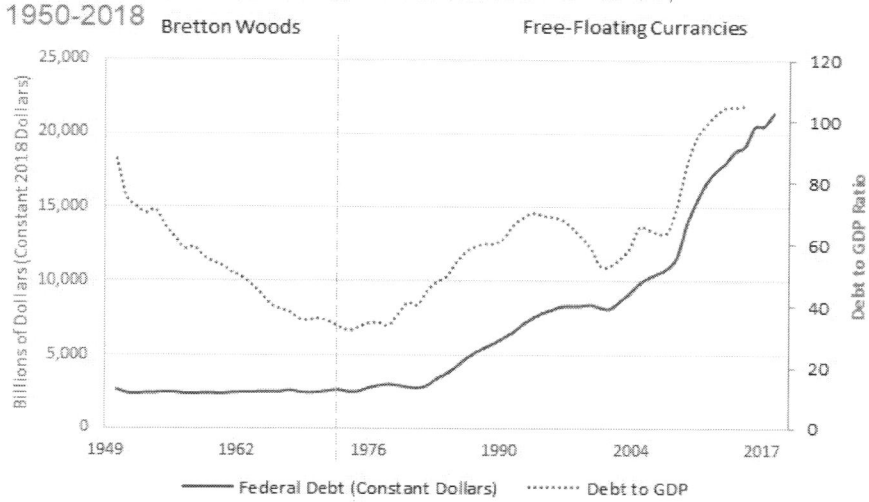

United States Federal Debt in Constant 2018 Dollars, 1950-2018

And this was not just the case for the United States, throughout the advanced economies – the G7 nations – the end of Bretton Woods marked the beginning of tremendous budget deficits. Keep in mind the scale of these deficits as well – for context the entire US defense budget is roughly three percent of current GDP, so a budget deficit of four or six percent is substantial.

Average Budget Deficit of the G7 Nations

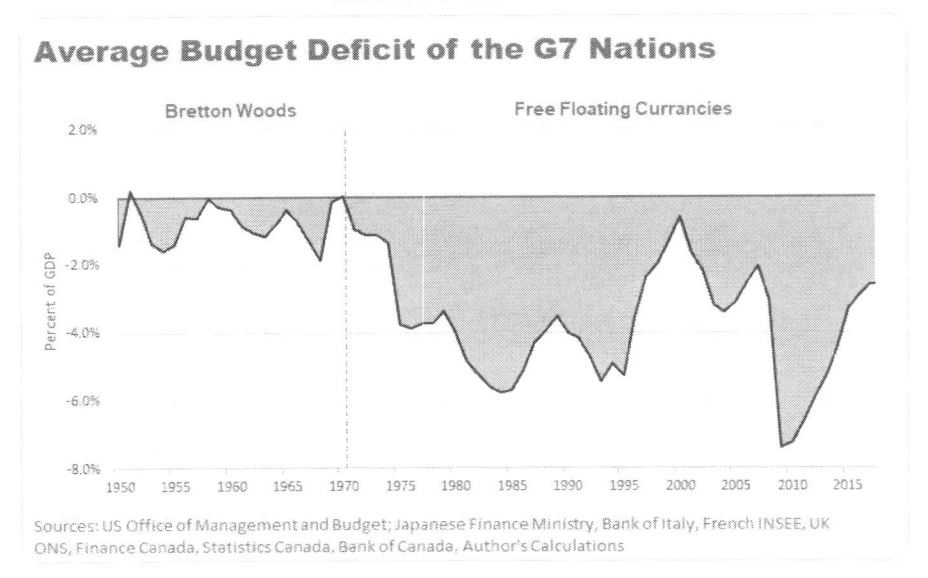

Sources: US Office of Management and Budget; Japanese Finance Ministry, Bank of Italy, French INSEE, UK ONS, Finance Canada, Statistics Canada, Bank of Canada, Author's Calculations

And naturally debt for all of the G7 nations has exploded – both as a percent of GDP and in total inflation adjusted terms.

Debt Increase for the G7 Nations

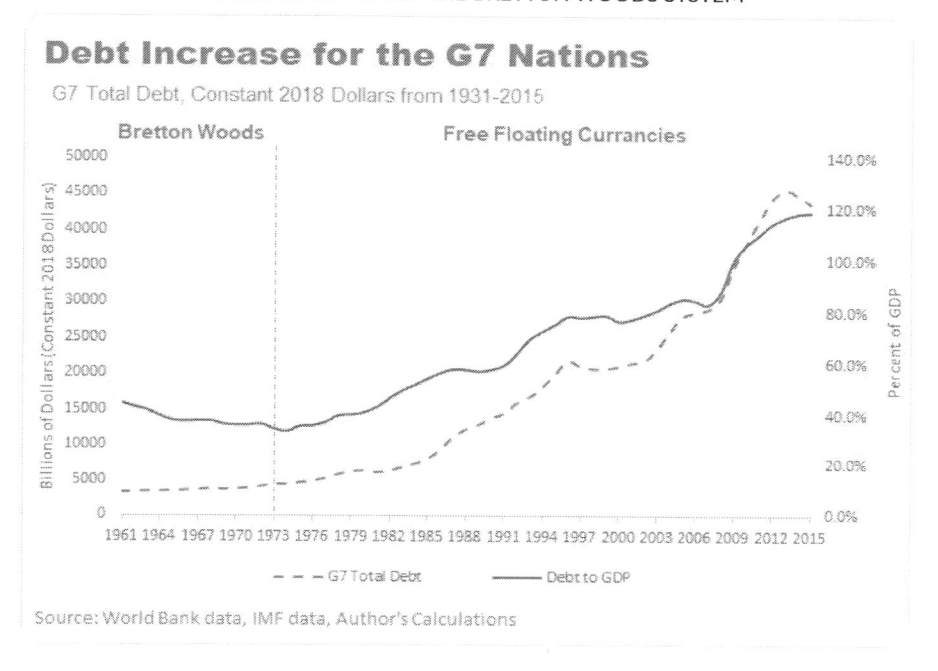

G7 Total Debt, Constant 2018 Dollars from 1931-2015

Source: World Bank data, IMF data, Author's Calculations

Free-Floating Currencies and Inflation

Although a slight tangent to the larger story of debt accumulation, there was another major impact of free-floating currencies which is worth exploring. With the end of the Bretton Woods system, nations began to devalue their currencies. Many nations did this to boost exports, or to stimulate their own economies, but there are several other consequences. This devaluation of currency helps to explain much of the rising inequality in the United States, as well as the rising costs for interest-rate-sensitive assets like education and housing. As the renowned economist John Maynard Keynes explained:

> ... By a continuing process of inflation, governments can confiscate, secretly and unobserved, an important part of the wealth of their citizens. By this method they not only confiscate, but they

confiscate arbitrarily; and, while the process impoverishes many, it actually enriches some. The sight of this arbitrary rearrangement of riches strikes not only at security, but at confidence in the equity of the existing distribution of wealth. Those to whom the system brings windfalls, beyond their deserts and even beyond their expectations or desires, become "profiteers," who are the object of the hatred of the bourgeoisie, whom the inflationism has impoverished, not less than of the proletariat[52].

A paper, *Monetary policy and the top one percent: Evidence from a century of modern economic history* by Mehdi El Herradi and Aurélien Leroy, lends empirical evidence to Keynes analysis. In their paper which analyzes the relationship between monetary policy and income inequality between 1920 and 2015 using annual data across 12 advanced economies, they note:

The last decades have been marked by a substantial rise in income and wealth inequality across the developed world. Low-income households in advanced economies have seen their wages stagnating, while wealth has never been so concentrated since the dawn of the 20th century[53].

They conclude based on their analysis that:

The results obtained from both empirical methods indicate that loose monetary conditions strongly increase the top one percent's income and vice versa. In fact, following an expansionary monetary policy shock, the share of national income held by the richest 1 percent increases by approximately 1 to 6 percentage points, according to estimates from the Panel VAR and Local Projections (LP). This effect is statistically significant in the medium run and economically considerable. We also demonstrate that the increase in top 1 percent's share is arguably the result of higher asset prices[54].

Looking at the Consumer Price Index (CPI) for the United States over the long-term shows the immediate and sustained inflationary impact of the free-floating monetary system. Most people today are used to the roughly 2% annual inflation - but until the current free-floating monetary system, such inflation only took place during wartime.

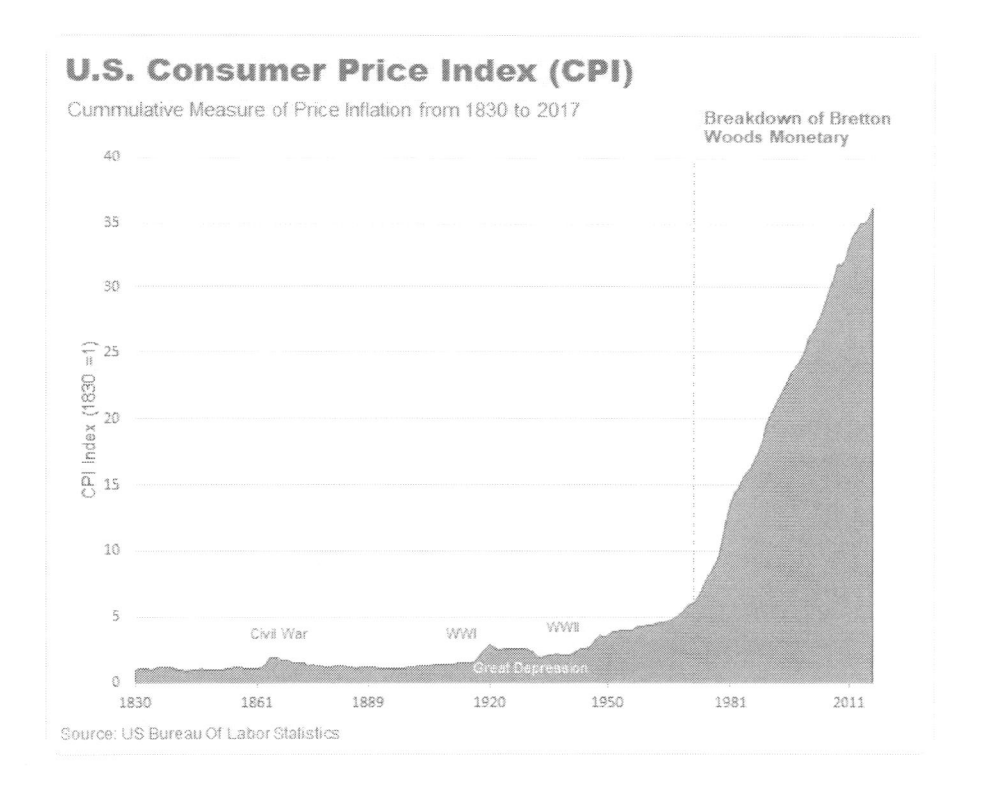

U.S. Consumer Price Index (CPI)
Cummulative Measure of Price Inflation from 1830 to 2017

Source: US Bureau Of Labor Statistics

This constant inflation distorts the prices and misallocates resources. But it also effectively functions as wealth tax, particularly on the poorest who are unable to buy appreciating assets like stocks. Since the end of the Bretton Woods system in 1971, the costs for interest-rate sensitive assets like education, healthcare, and housing has increased substantially – even in constant dollar values adjusted for inflation.

The Impact of Free-Floating Currencies

In summery, the end of the Bretton Woods system resulted in the emergence of free-floating currencies. These flexible exchange rates allowed nations to devalue their currencies which worked, along with other policies, to suppress consumption and thus boost exports – resulting massive trade imbalances which, combined with the demographic dividend, have moved growth to Asia (though an ultimately unsustainable way). These trade imbalances, and the absence of fiscal discipline under free-floating currencies, helped create enormous budget deficits and massive debt accumulation.

The sovereign debts that threatens the economic health of the globe were *enabled* by the demographic dividend of falling fertility rates and the GDP growth from increasing working-age populations, but they were *incentivized* by the changes to the global monetary system. The free-floating currencies which trapped nations in massive trade imbalances and removed the budget discipline imposed by fixed exchange rates further incentivizing budget deficits and debt accumulation.

III

Unsustainible Accumulation of Debt

The breakdown of the Bretton Woods system distorted the global economy. Free-floating currencies enabled artificially low interest rates and manipulated currency exchanges, which disconnected a vast amount of commerce from true supply and demand. But the largest and most unsustainable consequence of free floating currencies was the accumulation of debt by governments around the globe.

4

The Accumulation of Debt and Quantitative Easing

The Impact of a Sovereign Debt Crisis

While economists, think-tanks, and key institutions have been warning of the ultimate unsustainablity of the debts that have been accumulated for decades, the imminent transition of the Baby Boomer generation into retirees will be the trigger for a debt crisis fifty years in the making.

The risk to the world from the accumulated debt and the fall in the working-age population is more than just a question of possible cuts to promised entitlements or future tax hikes. Governments around the world have been running deficits for decades. Taking advantage of the large size of the *Baby Boomer* generation to create *pay as you go* systems that cannot be funded as the population of the developed world shifts from young to old. There will undoubtedly be cuts to promised entitlements and future tax hikes – but the real threat in the demographic decline lies in the leveraged nature of the world financial system and debt that nations have already taken on.

Financial institutions overvaluing subprime mortgages caused the 2008 financial crisis. Regarded as safe assets, they were highly leveraged in the banking system. When borrowers could not pay the mortgages, they fell in value. The leveraged nature of the assets meant that such loss of value created a deflationary spiral which wrecked havoc throughout the financial system. Havoc that affected not only homeowners, or banks which had made risky loans, but everyone in the economy. Similarly, but on a much greater scale, sovereign debt underpins the global financial system.

The nonpartisan Congressional Budget Office (CBO) has warned that the levels of debt accumulated by the United States:

> ... increase[s] the likelihood of a fiscal crisis, *an occurrence in which investors become unwilling to finance a government's borrowing unless they are compensated with very high interest rates*[55].

The Government Accountability Office (GAO) says that the nation is on an "Unsustainable Fiscal Path: Costs are Growing Faster than Revenues," and that "the nation's unsustainable fiscal path is not just something that we are predicting for the future – it is happening now[56]."

This not merely a problem for the United States, it is a global issue. In the U.K., the Office for Budget Responsibility (OBR) warns that the costs of an ageing population will send Government debt spiraling:

> *The national debt will more than triple from 85pc [percent] of gross domestic product (GDP) now to more than 280pc [percent] in 2067, the OBR said, as the surging costs of healthcare and pensions combine with a shrinking proportion of workers to pay for it all, creating an "unsustainable" burden*[57].

The IMF has warned Japan – The world's third largest economy and the nation with both the fastest ageing population and the largest debt to GDP ratio – that "Japan's public debt is unsustainable under current policies," IMF economists writing in *Defying Gravity: Can Japanese Sovereign Debt Continue to Increase Without a Crisis?* Warn that:

> *Almost all recent papers on Japanese government debt reach the same conclusion: the current course of fiscal debt dynamics is not sustainable... all find that without a drastic change in fiscal policy, the Japanese government debt to GDP ratio cannot be stabilized*[58].

And again, these are just a few of the countries that are facing unsustainable increases in sovereign debt. Nations of the world have borrowed enormous sums of money – an amount that they cannot possibly pay back in real terms. Globally sovereign debt has risen dramatically in the last fifty years.

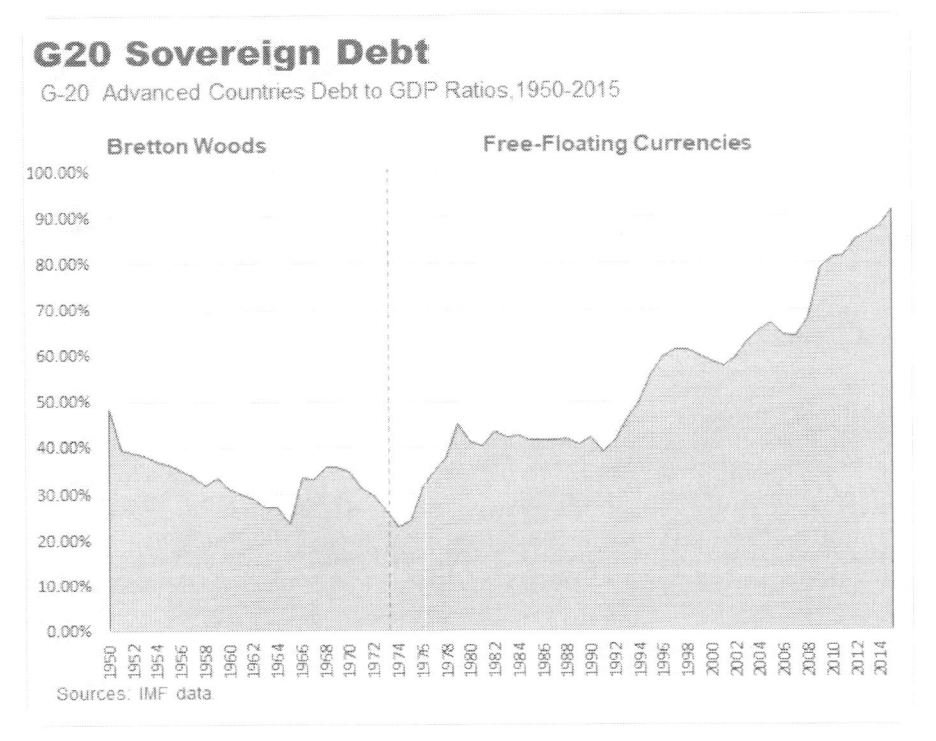

G20 Sovereign Debt
G-20 Advanced Countries Debt to GDP Ratios, 1950-2015

Bretton Woods | Free-Floating Currencies

Sources: IMF data

In the larger view of history, this is not unusual. As Reinhart and Rofoff have detailed in their seminal work *This Time is Different*, where they studied eight centuries of sovereign debt defaults and inflation crises, massive buildups of sovereign debts followed by defaults are not rare in a larger view of history. As they note:

> *All too often, periods of heavy borrowing can take place in a bubble and last for a surprisingly long time... This time may seem different, but all too often a deeper look shows it is not*[59];

The current economic situation of the world is unusual only because of its scale. Enabled by demographic factors and changes to the global monetary system. But the current debt accumulation is just the beginning. The debt buildup is projected to accelerate across the developed world - even without accounting for any potential economic downturns. The demographic shift from young to old means that spending on aging populations will rise as growth and tax revenues are projected to slow.

While official long-term debt projections are not available for most of the G-20 nations, the Congressional Budget Office in the United States and the Office for Budget Responsibility in the U.K. both publish long-term projections. The U.S. public debt for example, even using the CBO's fairly optimistic projections, will soon be almost double the debt accrued after World War II. Unlike in 1950, the debt has not accumulated from a brief but expensive war, but decades of unbalanced budgets which will be painful to reverse. Moreover, low economic growth and declining populations make the debt level much more unsustainable than in 1950. The United Kingdom has a very similar projection. You can see that both nations debts surged after the 2008 financial crisis and are projected to rise significantly in the coming decades because of the costs of ageing populations.

U.S. and U.K Debt to GDP

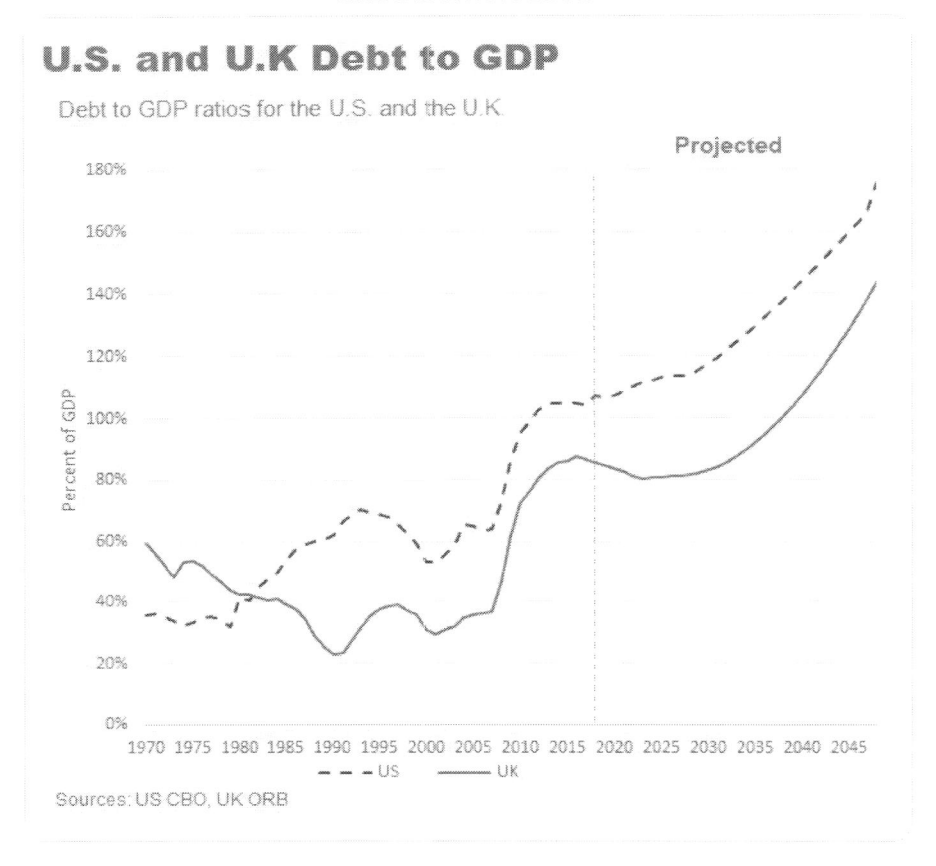

Debt to GDP ratios for the U.S. and the U.K.

Sources: US CBO, UK ORB

Looking at the United States, its clear that in fact, while government debt declined as a percentage of GDP after World War II – the debt never declined in real terms. The economy just grew. The debt was never paid down, it stayed steady in real terms while the economy grew larger.

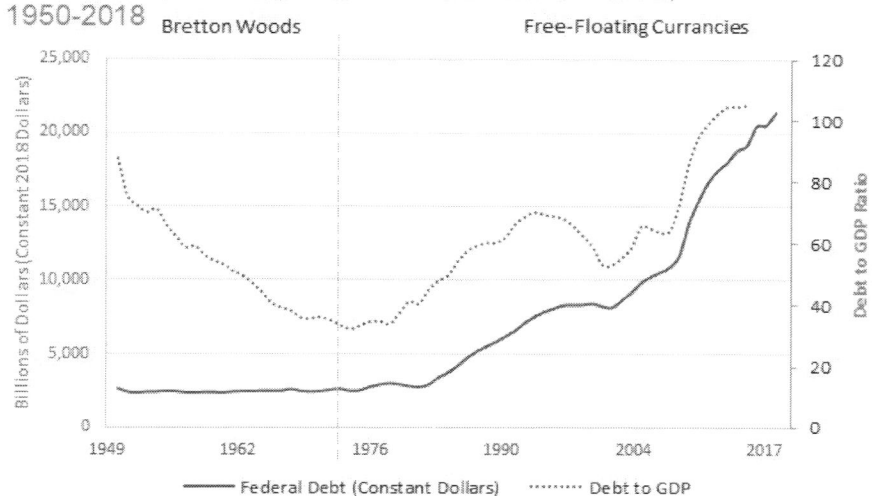

United States Federal Debt in Constant 2018 Dollars, 1950-2018

However, no one believes that we are likely to see the same ten to fifteen percent growth rates that we saw in the 1950s and 1960s. The world is aging – and even in the US with relatively healthy demographics, growth is expected to slow even more.

Writing in *The Future of Public Debt: Prospects and Implications*, economists Stephen G Cecchetti, M S Mohanty and Fabrizio Zampolli argue that the debts accumulated by the developed world are already unsustainable:

> *According to the OECD, total industrialized country public sector debt is now expected to exceed 100% of GDP in 2011 – something that has never happened before in peacetime. As bad as these fiscal problems may appear, relying solely on these official figures is almost certainly very misleading. Rapidly ageing populations present a number of countries with the prospect of enormous*

future costs that are not wholly recognized in current budget projections.

Today, interest rates are exceptionally low and the growth outlook for advanced economies is modest at best. This leads us to conclude that the question is when markets will start putting pressure on governments, not if. When, in the absence of fiscal actions, will investors start demanding a much higher compensation for the risk of holding the increasingly large amounts of public debt...

It is clear that sovereign debt has already begun to lose at least some amount of confidence. In the *OECD Sovereign Borrowing Outlook 2018*, the OECD reports that:

A number of countries have been downgraded by the three big credit agencies during the past decade – in effect shrinking the pool of government bonds in the prime category to 11, down from 19 a decade ago. Notably, Ireland lost its AAA rating status in 2009, Spain in 2010, the United States in 2011(only by Standard and Poor's), Austria and France in 2012, the United Kingdom in 2013, and Finland in 2014. More broadly, credit ratings of many countries have steadily shifted down since the GFC [Great Financial Crisis][60].

The report also shows a very interesting graphic, reproduced below with permission, illustrating how in 2008, almost 90% of sovereign debt was issued at the highest possible rating - as of December 2017 only 30% of sovereign debt issued was rated AAA. In fact, while in 2008 only about 10% of debt was below the *Prime* and *High* ratings level, today roughly half of sovereign debt issued is *Upper Medium* grade or below.

THE ACCUMULATION OF DEBT AND QUANTITATIVE EASING

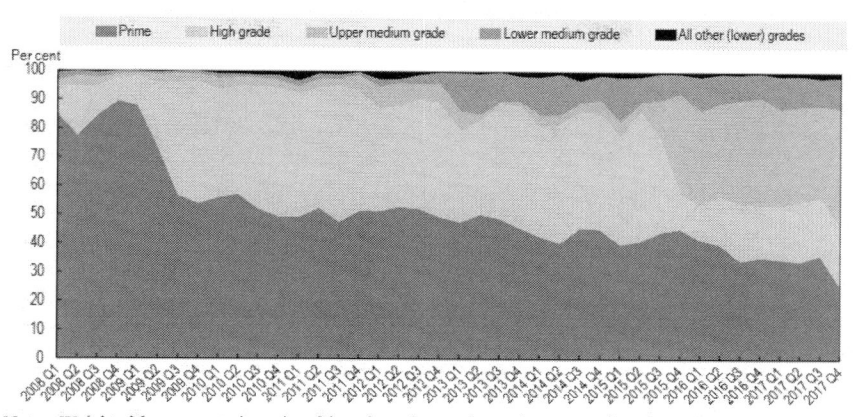

Notes: Weighted by amounts issued and based on the maximum issuance rating from three rating agencies: Fitch, Moody's and Standard and Poor's.

Keep in mind that Greece's credit rating was in the Upper Medium grade as of January 2009 when they had an *A* rating.

Why We Haven't Seen A Crisis Yet

Despite the growing debt and worrying future projections of more debt, there have been few signs of concern from the markets. Interest rates on sovereign debt have not risen as credit ratings have fallen, in fact, interest rates are at all time lows. It is these low interest rates that have allowed the world to accumulate such debt without a crisis.

Bond Yields for Selected Nations, 1971-2019

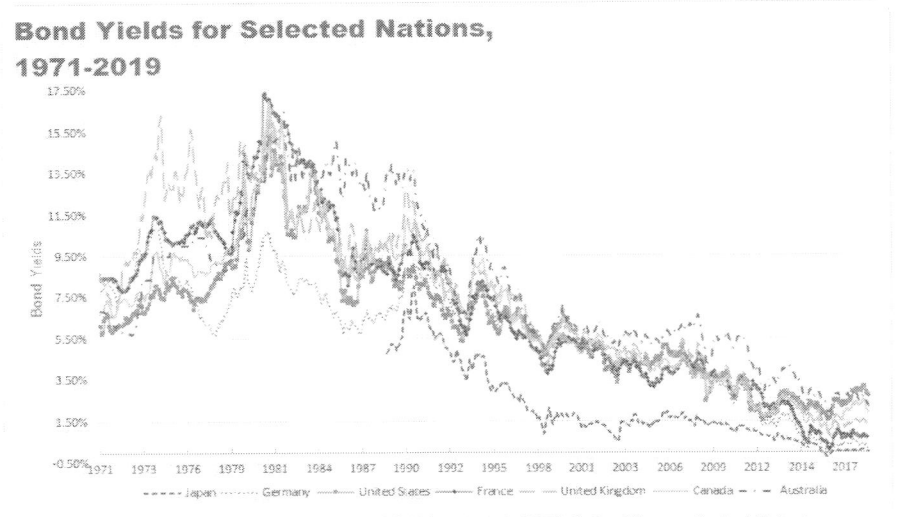

Source: Organization for Economic Co-operation and Development, via FRED, Federal Reserve Bank of St. Louis

Many studies have shown the strong impact of the age structure on long-term interest rates – *such as: Demographics and The Behavior of Interest Rates 2015.* As a greater proportion of the population is in their peak earning years and saving for retirement this increases the total amount of savings available for investment. This increased supply of savings will tend to lower interest rates and has been one of the major reasons that rates have been pushed down – especially in inflation adjusted terms. As the worldwide Baby Boomer generation saves more and more for retirement, they invest more and more in assets like government bonds.

In the 2016 Federal Reserve Bank of San Francisco Working Paper, *Demographics and Real Interest Rates: Inspecting the Mechanism*, authored by Carlos Carvalho, Andrea Ferrero, and Fernanda Nechio, the economists write that:

> *The Demographic developments that most advanced economies*

are undergoing are a natural explanation for the prolonged decline of global real interest rates. The main channel through which demographics affect the real interest rate is the increase in life expectancy. At all stages of their life cycle, individuals save more to finance consumption over a longer time horizon. In our calibrated model, the demographic transition can account for about one third to a half of the overall decline in the real interest rate since 1990[61]

But demographic factors alone do not explain today's ultra-low interest rates. Where Greece pays a lower interest rate than the United States, or where Japan pays negative interest although they have the highest debt to GDP ratio in the world - they are a product of quantitative easing.

Quantitative Easing

In 2001 the Bank of Japan (BOJ) was forced to begin *ryōteki kin'yū kanwa (quantitative easing)* in 2001. This *quantitative easing* policy, or simply QE, was adopted by the rest of the G7 nations following the 2008 financial crisis when they too were forced to take on massive debts to bailout their financial systems. The quantitative easing engaged in by the major central banks has worked to push down long-term interest rates to unprecedented lows.

There is a great deal of misunderstanding about quantitative easing. Some prominent investors characterized it as *printing money* and expected it to cause massive inflation. Others have used the lack of inflation as an argument about why sovereign debt doesn't matter. Neither of these characterizations are accurate. Ultimately QE is neutral to the monetary base and does not remove government liabilities –although it does lower debt servicing costs. Former Federal Reserve chair Ben Bernanke has attempted to refute the idea that QE is *printing money* explaining that:

> ...sometimes you hear the Fed is printing money, that's not really happening, the amount of cash in circulation is not changing. What's happening is that banks are holding more and more reserves with the Fed[62].

Essentially, the mechanism of QE is that the Federal Reserve and other central banks, are buying government bonds – but not directly from the Treasury Department. They're buying government bonds from banks and paying for them with central bank reserves.
Crucially, these cash reserves are held by the central bank and pay an interest rate just as the treasury bonds did. In a sense the Federal Reserve swapped ten year treasury bonds with hypothetical *Federal Reserve Bonds*. That's why you can see a graph of the money supply that shows a massive increase, and yet inflation has not spiked. The money never entered the economy – it is being held in a vault at the central bank (or more realistically is just a number in a spreadsheet). In practical terms it was an asset swap, not a true increase in the money supply. The graphic below shows how QE affects the balance sheets of the banking system, the central bank, and the Treasury in a hypothetical QE process.

Quantiative Easing Process

Bank			Central Bank			Treasuary	
Assets	Liabilites		Assets	Liabilites		Assets	Liabilites
Reserves: $ 50	Deposits: $ 60		T-Bonds: $ 50	Reserves: $ 50		Money $ 75	T-Bonds: $ 75
Loans: $150							
T-Bonds: $ 25							

Bank			Central Bank			Treasuary	
Assets	Liabilites		Assets	Liabilites		Assets	Liabilites
Reserves: $ 75	Deposits: $ 60		T-Bonds: $ 75	Reserves: $ 75		Money $ 75	T-Bonds: $ 75
Loans: $150							
T-Bonds: $0							

QE is essentially an asset swap where the amount of money in circulation remains unchanged. It does not increase or decrease the money supply directly. And neither does it reduce the fundamental debt burden and obligations of governments. For as part of the QE process, the central banks pays the bank interest on excess reserves (IOER) they hold with the central bank. An interest rate typically very close to the yield on the bonds themselves.

Economists for the Peterson Institute for International Economics and the Federal Reserve Bank of New York explain how this *asset swap* impacts long-term interest rates. in the 2010 paper, *The Financial Market Effects of the Federal Reserve's Large-Scale Asset Purchases*, they describe how:

> *The primary channel through which LSAPs [Large Asset Purchases – i.e. QE] appear to work is by affecting the risk premium on the asset being purchased. By purchasing a particular asset, a central bank reduces the amount of the security that the private sector holds, displacing some investors and reducing the holdings of others, while simultaneously increasing the amount of short-term, risk-free bank reserves held by the private sector. In order*

for investors to be willing to make those adjustments, the expected return on the purchased security has to fall. **Put differently, the purchases bid up the price of the asset and hence lower its yield.** *This pattern was described by Tobin (1958, 1969) and is commonly known as the "portfolio balance" effect.*[63] *[emphasis added]*

Put more simply, by lowering the total amount of Treasury Bonds available in the private sector the central bank has reduced the amount of safe assets available to investors. With supply reduced and demand unchanged investors will bid the interest rates on government bonds lower. This *portfolio balancing* effect also impacts other securities. Investors searching for yield will also bid interest rates lower on alternative investments such has highly rated corporate debt etc.

The Inherent Limits of QE

Importantly though, this QE process is only possible as long as as there are bonds being held by banks. Pension funds or other investors are not eligible to keep reserves at the central bank, and of course banks hold a finite amount of government bonds. Therefore QE cannot be continued indefinitely.

Already the BOJ has begun to run out of eligible bonds to buy – or swap really – for reserves at the central bank. As Ben Bernanke noted in late 2016 "...constraints on the availability of JGBs [Japanese Government Bonds] were seen in many quarters as limiting the BOJ's ability to maintain its easy policies beyond the next year or two[64]." In September 2016, the *Financial Times* raised the question "When will the ECB run out of bonds to buy? Warning that the "QE programme could hit a wall as early as this year because of lack of eligible debt".[65]

If the BOJ or the ECB was to simply buy bonds directly from the treasury without removing an equivalent amount from circulation by holding it as reserves it would be massively inflationary. Because of this, QE is an inherently temporary solution. Ultimately, QE can only delay the inevitable rise in the yields on sovereign debt.

Now, while the BOJ has just about purchased every available JGB, the ECB is only facing the self-imposed political limit of 30% of outstanding bonds. This of course could be changed at any time and the QE continued. In a 2016 European Parliament report, *Limits in terms of eligible collateral and policy risks of an extension of the ECB's quantitative easing programme*, economists Jens Boysen-Hogrefe, Salomon Fiedler, Nils Jannsen, Stefan Kooths and Stefan Reitz acknowledge the inherent limits to a QE program - saying that the ECB will be able to continue for another year , according to announced plans, but that:

> *Overall, the ECB probably must adjust the limits and criteria for eligible assets if the QE programme is to be extended further[than March 2017]. There are several ways how the ECB could adjust these limits and criteria to significantly increase the amount of eligible assets. However, all of these adjustments would involve the following drawbacks: financial risks would significantly rise, questions of monetary financing would intensify, or market functioning would be put at risk. Moreover, if the ECB continuously changes the limits and criteria this could raise questions whether these limits and criteria have been chosen rather arbitrarily*[66].

So while Japan is reaching the ultimate limits of QE Europe is only approaching a self-imposed limit. But the point remains that QE is an inherently temporary solution.

The Economic Distortion of Artificially low Interest Rates

Now, while QE has lowered long-term interest-rates and avoided a debt crisis it has not come without significant drawbacks. The economists from the 2016 European Parliament report also note the risks inherent in the QE program's attempt to lower interest rates saying:

> ...very low interest rates for an extended period of time stimulate risk-taking (Rajan 2005), potentially fuels asset price bubbles, in turn increasing systemic risks and possibly triggering banking crises. These risks of very expansionary monetary policy tend to increase the longer it is in place (Maddaloni and Peydro 2011,2012). Moreover, very expansionary monetary policy can also trigger the misallocation of real resources and thereby dampen potential growth (White 2012) and hinder necessary adjustment processes in the aftermath of financial crises (Hoshi and Kashyap 2004;Caballero et al. 2008)[67].

And indeed, while QE has lowered interest rates on sovereign debt and prevented a fiscal crisis up till now, it has come at the cost of distorting the global economy even more than before the 2008 financial crisis - delaying but worsening an inevitable reckoning.

Zombie Economy

One of the most significant economic developments since the Great Recession, and a consequence of ultra-low interest rates, has been the zombification of the economy.

A *zombie company* is a term introduced to the lexicon by an influential paper, *Zombie Lending and Depressed Restructuring in Japan,* by economists Ricardo J. Caballero, Takeo Hoshi, and Anil K Kashyap.

The official definition of a *zombie company* according to the Bank for International Settlements (BIS) is "is a publicly traded firm that's 10 years or older with a ratio of earnings before interest and taxes (EBIT) to interest expenses of below one." More simply put, zombie companies are companies that are unprofitable – so unprofitable they are unable to pay even the interest on their debt out of their profits. They are effectively bankrupt but *kept alive* by banks continuing to lend them money to pay their existing loans.

This phenomenon first began in Japan after their real estate and stock market bubble popped in the early 1990s. The economists writing *Zombie Lending and Depressed Restructuring in Japan* have proposed that zombie companies are to blame for Japan's Lost Decade(s), writing:

> *We propose a bank-based explanation for the decade-long Japanese slowdown following the asset price collapse in the early 1990s...Large Japanese banks often engaged in sham loan restructurings that kept credit flowing to otherwise insolvent borrowers (which we call zombies). We examine the implications of suppressing the normal competitive process whereby the zombies would shed workers and lose market share. The congestion created by the zombies reduces the profits for healthy firms, which discourages their entry and investment. We confirm that zombie-dominated industries exhibit more depressed job creation and destruction, and lower productivity*[68].

Following the Great Recession, zombie companies became a worldwide phenomenon. Even with today's very low interest rates; more and more companies are unable to pay the interest on their debts out of profits. According to the BIS, the share of zombie companies in the US **doubled** between 2007 and 2015, rising to around 10% of all public companies. And counterintuitively, as interest rates have fallen lower and lower the number of zombie companies has increased. Economists

Ryan Niladr Banerjee and Boris Hofmann, writing in the BIS quarterly review, describes this seemingly paradoxical result:

> Using firm-level data on listed firms in 14 advanced economies, we document a ratcheting-up in the prevalence of zombies since the late 1980s. Our analysis suggests that this increase is linked to reduced financial pressure, which in turn seems to reflect in part the effects of lower interest rates. We further find that zombies weigh on economic performance because they are less productive and because their presence lowers investment in and employment at more productive firms[69].

It part this may be because low interest rates signify a weak banking system. Banks may be reluctant to allow a company to fail – even if there is little hope of eventual repayment – because it would be too painful to accept the losses on the loans already lent to these companies. And of course the ultra-low interest rates created by central banks *unconventional monetary policy* since 2008 keeps the costs of servicing debt low.

These studies likely understate the problem of zombie companies for the economy. Corporate leverage has surged in the last three years since the BIS analysis - and a company doesn't have to be as far gone as a zombie to be at risk of default if interest rates rise. As economist Daniel Lacalle writes:

> At the end of the day, 10.5% means that 89.5% are not zombies. But that analysis would be too complacent. According to Moody's and Standard and Poor's, debt repayment capacity has broadly weakened globally despite ultra-low rates and ample liquidity. Furthermore, the BIS only analyses listed zombie companies, but in the OECD 90% of the companies are SMEs (Small and Medium Enterprises), and a large proportion of these smaller

non-listed companies, are still loss-making. In the Eurozone, the ECB estimates that around 30% of SMEs are still in the red and the figures are smaller, but not massively dissimilar in the US, estimated at 20%, and the UK, close to 25%[70].

Indeed, corporate debt has risen substantially over the last fifty years, and is now above the levels seen before the 2008 crisis in the United States.

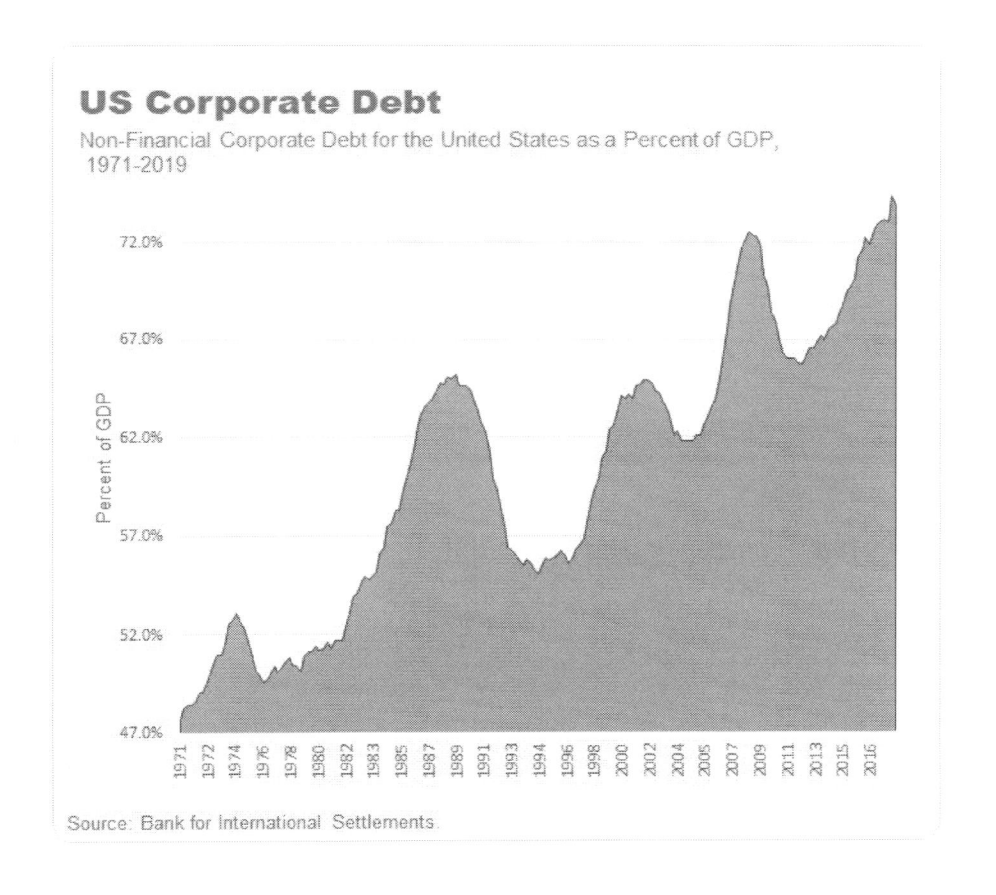

US Corporate Debt
Non-Financial Corporate Debt for the United States as a Percent of GDP, 1971-2019

Source: Bank for International Settlements.

In a market economy, resources are allocated according to profitably – this allows resources to flow to where they best utilized. Keeping companies that are unprofitable alive misallocates resources and, as numerous studies have shown, slows the growth of the entire economy. Potentially leading to stagnation as has been seen in Japan sine the 1990s.

Zombie companies pose a significant challenge for central banks, because in a very real sense their hands are tied and they cannot raise interest rates significantly without causing mass bankruptcies. The world now has the impossible choice of permanently reduced productivity and slower economic growth - or the mass bankruptcy of a significant percentage of the economy.

Pensions and Low Interest Rates

Another time bomb created by artificially low-interest rates, lies in pension funds. Ultra-low interest rates have forced pension funds to take on riskier assets to maintain yields. in 1962, pensions funds projected an average annual return of 8% and the yield on 30-year treasury bonds was just under 8%. Today, the average pension fund projects an annual return of just under 8%, while the 30-year treasury rate has fallen to less then 2%. In order to meet the required returns, pension funds have thus been pushed into equities and other riskier assets. In 1962 roughly 95% of pension fund assets were in fixed-income and cash. As of 2012, only 25% of pension fund assets are in fixed-income assets or cash - with the remaining 75% in equities and other higher yield alternatives. According the a report from the PEW charitable trust:

> *In a bid to boost investment returns, public pension plans in the past several decades have shifted funds away from fixed-income investments such as government and high-quality corporate bonds. During the 1980s and 1990s, plans significantly increased their reliance on stocks, also known as equities. And during*

the past decade, funds have increasingly turned to alternative investments such as private equity, hedge funds, real estate, and commodities to achieve their target investment returns[71].

Because of this shift towards riskier assets, according to researchers from the Pew Charitable Trusts in a 2018 paper: "Public pension systems may be more vulnerable to an economic downturn than they have ever been[72]."

Accounting for Risk in Public Pension Funds

There is one other key reason public pension funds have transitioned to riskier assets. States have passed laws exempting state government pension plans from the standards that private pension plans are held to. These public pension plans are permitted to use generous assumptions about risk that are not permitted in the private sector. So while public pensions have moved into objectively riskier assets, they haven't been forced to account for that risk.

According to a 2018 report, *Unaccountable and Unaffordable*, by the American Legislative Exchange Council, If public sector pension plans were held to the same standards as the private sector, even with their extremely optimistic estimates, they would be criminally underfunded. As the report states:

> *However, the Pension Protection Act does not apply to public sector DB pension plans. Using the states' own estimates of their liabilities and assets, 32 states are at risk of default by private sector standards. If the Pension Protection Act were applied to the public sector and states had to use a similar discount rate as the private sector, about 4.5 percent, only Wisconsin's pension system has enough assets to be considered stable*[73].

And this is not a controversial point – an overwhelming majority of leading economists agree that the government accounting standards used by US state and local governments understate their pension liabilities and the true cost of future pensions.

A recent estimation from 2018 calculates that "Unfunded liabilities of state-administered pension plans, using a proper, risk-free discount rate, now total over $5.96 trillion. The average state pension plan is funded at a mere 35 percent."

Sovereign Debt by Country

With the aging of the world the tailwinds of growing economies and low interest rates that have allowed debt to expand have shifted. Additionally, the geopolitical environment is becoming increasingly volatile, and trade imbalances and distortion in the global economy caused by free-floating currencies have begun to be a real issue. It is clear that the forces which have enabled this buildup of debt have or are in the process of reversing.

Moreover, QE cannot be continued indefinitely. And already low interest rates have restructured the entire world economy in a way that is ultimately unsustainable. Virtually all nations have taken on enormous debts - possible only because of the ultra-low rates. In order to fully understand the buildup of sovereign debt around they world we need to look at some on the key economics in greater depth.

5

Japan

The Plaza Accords

In 1985, West Germany, France, the United States, Britain, and Japan signed the Plaza Accords. With the collapse of the Bretton Woods system fourteen years earlier, there was no longer an automatic mechanism to keep imports and exports between nations in balance – and suddenly currencies being under or overvalued relative to each other mattered. The United States began to see a widening trade deficit, as other nations, particularly Japan, devalued their currencies to indirectly boost exports and stimulate economic growth. Unhappy with the situation, the United States negotiated an agreement with these nations to allow their currencies to strengthen against the dollar.

The finance ministers of (from left) West Germany, France, the United States, Britain and Japan signed a historic currency adjustment agreement in New York City's Plaza Hotel in 1985.

As Japan allowed their currency to appreciate, they were immediately thrown into a recession. Germany's currency also strengthened following the Plaza Accords, and caused a brief recession there. But the German central bank was cautious in its stimulus. Japan however, embraced the possibilities enabled by the free-floating monetary system. They massively stimulated their economy by lowering interest rates and ultimately created a titanic asset bubble in real estate and equities.

Nikkei 225 Index - Adjusted for Inflation
Yen (Constant 1956 Dollars)
8000
7000
6000
5000
4000
3000
2000
1000
0
Japanese Stock Bubble
1957 1961 1965 1969 1973 1978 1982 1986 1990 1994 1998 2002 2006 2010 2014 2019
Source: www.macrotrends.net and author's calculations

When this bubble popped in 1991, their stock market and real estate values crashed. Leaving highly leveraged Japanese banks and insurance companies with massive amounts of bad debt. Understandably, to avoid the pain of a deep recession, Japan bailed out their banks and insurance companies. Avoiding a deeper recession - but also avoiding correcting the mis-allocation of resources that had taken place during the inflationary boom. This led to decades of stagnate growth known as *Ushinawareta Jūnen – the* Lost *Decades.*

* * *

Japanese Sovereign Debt

Since the beginning of Japan's *Lost Decades*, the country has been mired in debt – with the highest debt to GDP ratio of any nation. Additionally, Japan's *Baby Boomer* generation started ageing before most other countries. And their fertility rates fell to particularly low levels. Japan not only has the world's worst debt ratio – but they also have the oldest population as well.

However, Japan has had several advantages to deal with their debt. Firstly, Japan's debt expanded in an era when the rest of the world's population was in their prime earning and saving years. Interest rates continued to fall worldwide. Japan's debt is held almost entirely by Japanese citizens and institutions. Compared to other nations, the fraction of Japan's debt held by foreigners is very low. This reduces the risk that foreign investors will lose confidence in Japanese Government Bonds (JGBs) and try to move their money out of the country.

JAPAN

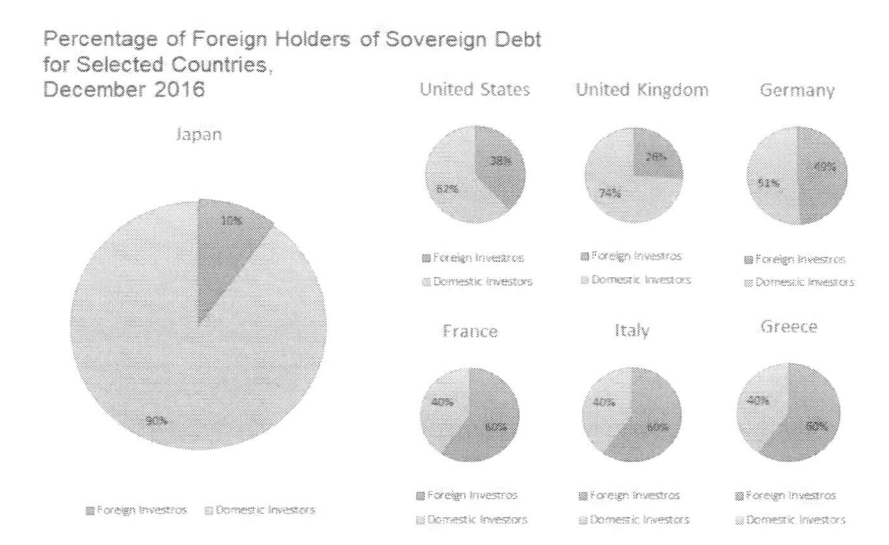

Percentage of Foreign Holders of Sovereign Debt for Selected Countries, December 2016

Japan: Foreign Investors 10%, Domestic Investors 90%
United States: Foreign Investors 38%, Domestic Investors 62%
United Kingdom: Foreign Investors 26%, Domestic Investors 74%
Germany: Foreign Investors 49%, Domestic Investors 51%
France: Foreign Investors 40%, Domestic Investors 60%
Italy: Foreign Investors 40%, Domestic Investors 60%
Greece: Foreign Investors 40%, Domestic Investors 60%

Japan built up this debt over the decades by years of deficit spending – over 8.6% of GDP in 2012 for example. Looking at their 2017 budget, an incredible thirty-five percent of revenues came from issuing new bonds.

Japan's 2017 Budget, Total Revenues

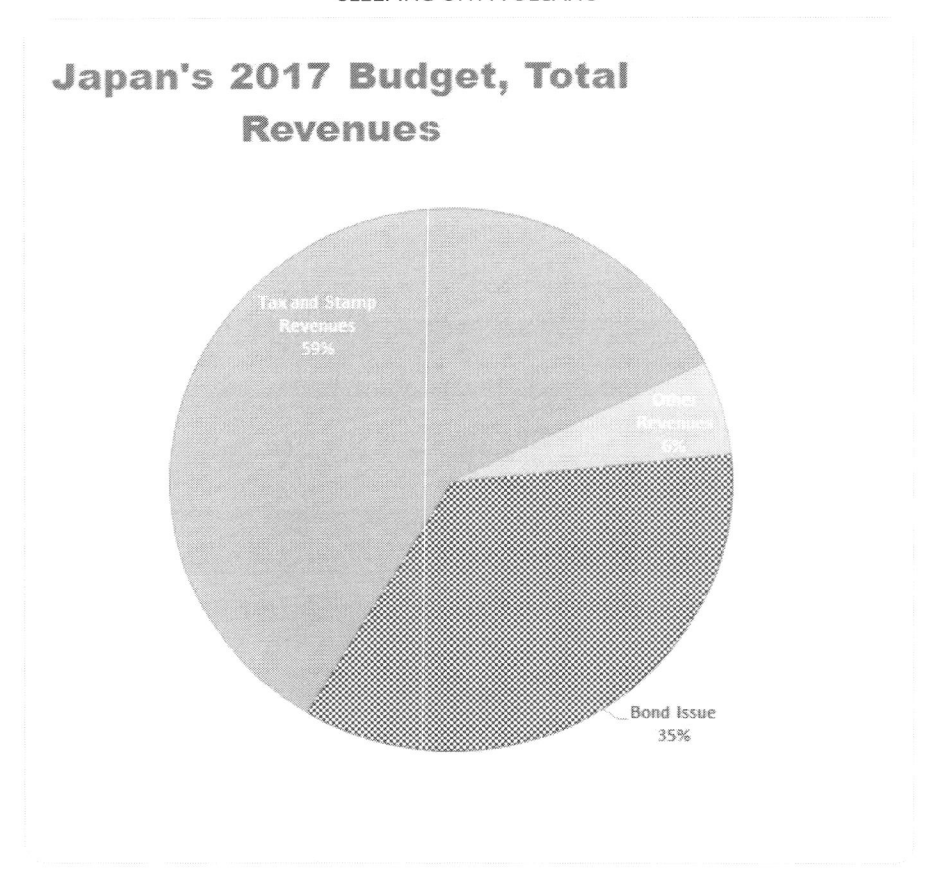

- Tax and Stamp Revenues 59%
- Bond Issue 35%
- Other Revenues 6%

Even more alarming, in the same year some 24% of government spending was just to service Japan's already existing debt.

Japan's 2017 Budget, Total Expenditures

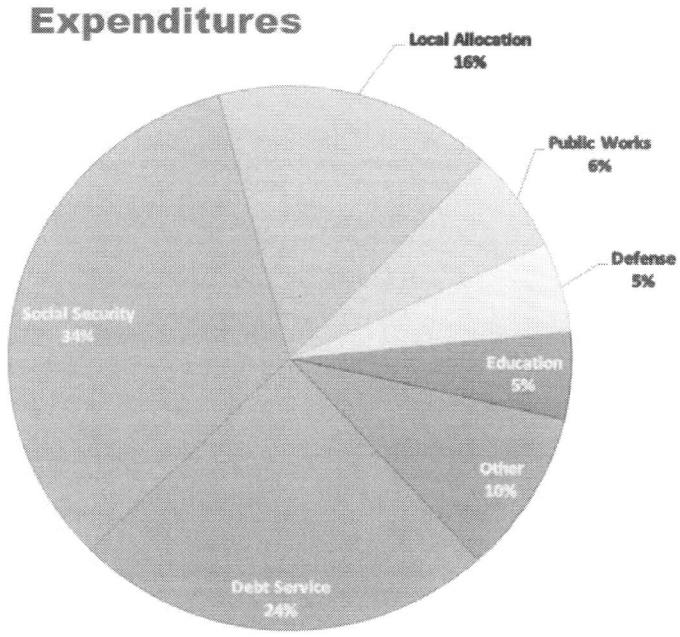

- Local Allocation 16%
- Public Works 6%
- Defense 5%
- Education 5%
- Other 10%
- Debt Service 24%
- Social Security 34%

And 2017 was not an outlier. If anything, 2017 was one of the better recent years. Japan has been running large – sometimes incredibly large– deficits since 1973 when Japan's currency began to float.

Japanese Spending and Revenue

Japan Spending and Revenue as a Percent of GDP, from 1965-2017

Source: Japanese Ministry of Finance

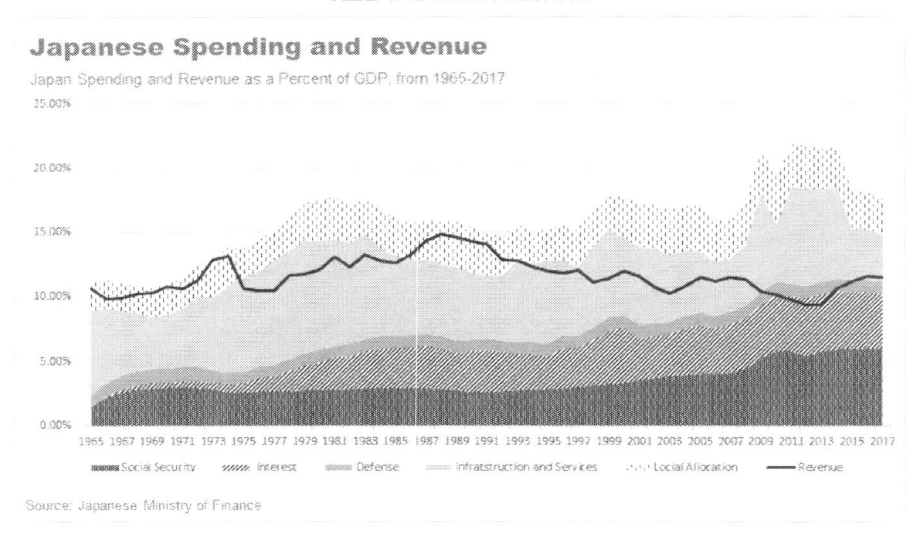

This massive deficit spending has taken the Japanese government from almost no debt in 1970 to almost 250% of debt to GDP today.

G7 Nations Debt

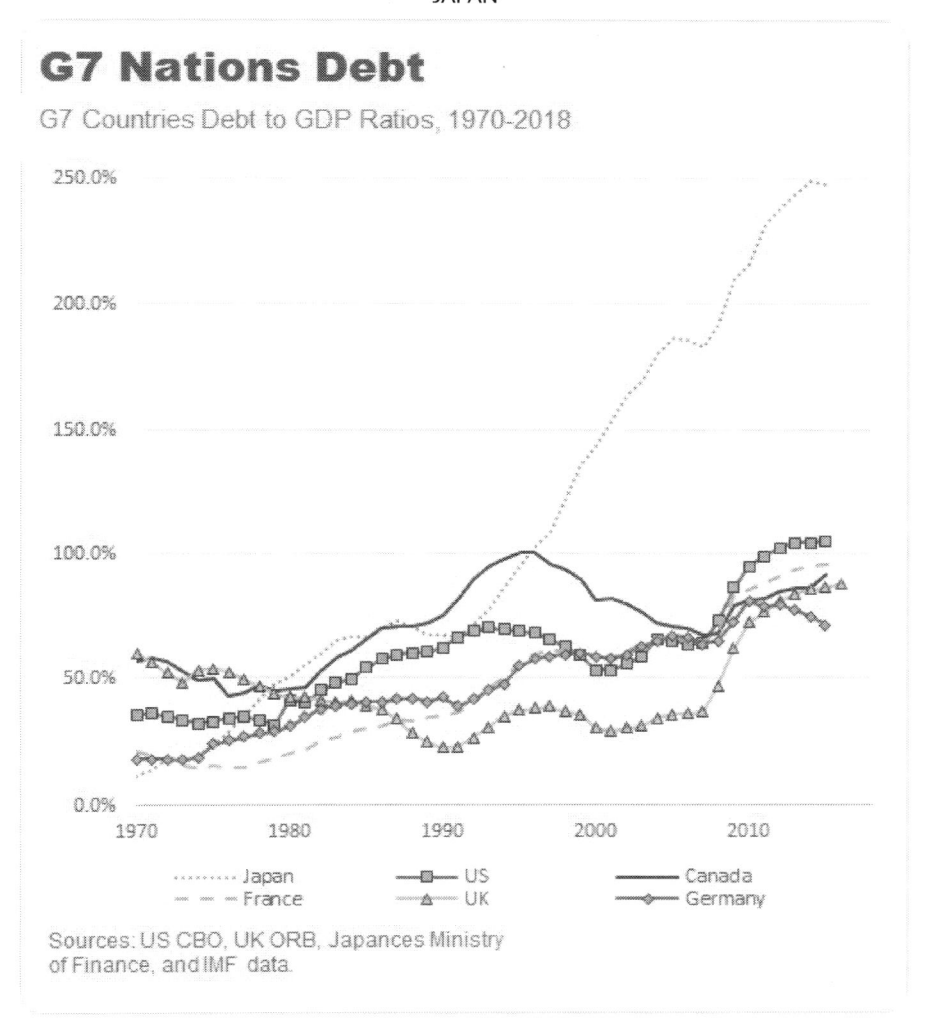

G7 Countries Debt to GDP Ratios, 1970-2018

Sources: US CBO, UK ORB, Japances Ministry of Finance, and IMF data.

The Japanese Ministry of Finance projections show that even under their fairly optimistic assumptions – including a 10% consumption tax hike which has since been delayed – they will continue with substantial deficits for the foreseeable future.

Japanese Spending and Revenue (Projected)

Japan Spending and Revenue as a Percent of GDP,
1965-2027, 2018-2027 Projected (Projections Assume 10% Increase in Comsumption Tax - Since Delayed)

Source: Japanese Ministry of Finance

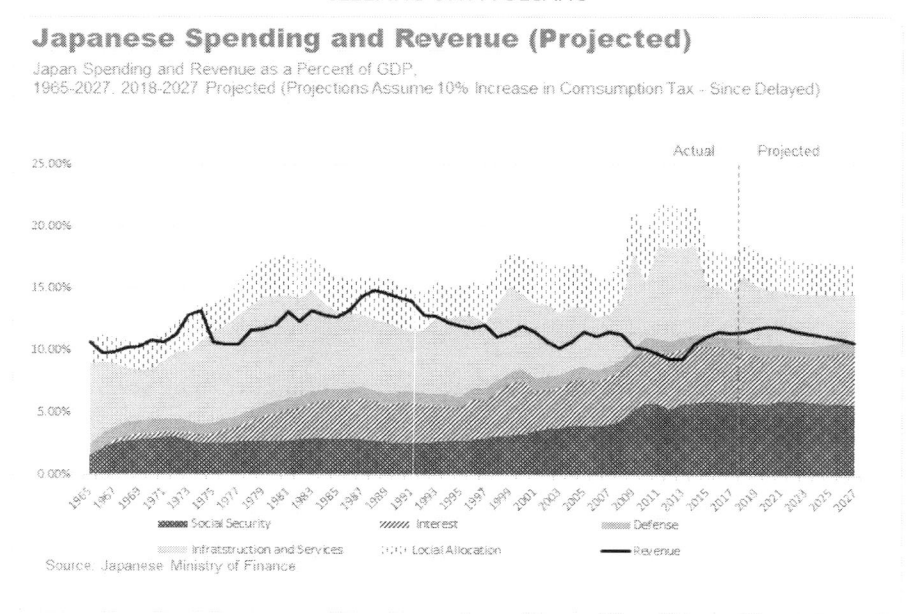

Consensus of Unsustainablity

As the Japanese population as aged, their costs for pensions and healthcare have risen, while their tax revenues have fallen. IMF economists writing in *Defying Gravity: Can Japanese sovereign debt continue to increase without a crisis?* Warn that:

> Almost all recent papers on Japanese government debt reach the same conclusion: the current course of fiscal debt dynamics is not sustainable... all find that without a drastic change in fiscal policy, the Japanese government debt to GDP ratio cannot be stabilized[74].

Japan's own cabinet office when asking the question – "is the Japanese Fiscal Situation Sustainable?" concluded that:

Formal tests report that sustainability has been lost since late 1990s and that Medium-term simulations show that without policy action debt to GDP ratio continues to rise. Researchers concerned with sustainability unanimously agree that the Japanese government will have to generate sufficient fiscal surplus for achieving sustainability. Moreover, when the share of foreign investors increases, as indicated by Tokuoka (2010) and Hoshi and Ito (2012), the surge in JGB bond yields is apprehended, and it may induce the crowding - out of private capital formation or reduce non - interest spending owing to the increase in interest rate payment. In addition, following the results of Onji et al. (2012), the JGB market may become volatile once government financial institutions quit their role of primary purchasers of JGBs. This could also prove detrimental to the macro economy as well as the government budget[75]

Or the 2011 National Bureau of Economic Research paper *Japanese Government Debt and Sustainability of Fiscal Policy* where Takero Dio, Takeo Hoshi, and Tatsuyosho Okimoto analysis Japan's debt and conclude that:

> *...the Japanese fiscal policy can be sustainable if Japan can go back to the policy regime in 1885-1925 (roughly under the gold standard) and in 1950-1970 (effective balanced budget legislation) [the Bretton woods system] with high probability. Without such drastic changes, the fiscal policy is unsustainable.*

> *...All the results point to the same conclusion: the Japanese government debt poses serious challenges. To stabilize the debt to GDP ratio, Japan needs to implement a tax rate hike with an extraordinary magnitude. Such tax increase to make the fiscal policy sustainable would represent a drastic departure from the Japanese fiscal policy in the last 30 years. The fiscal policy in Japan*

is found to be unsustainable even when we allow the possibility of regime changes. If the government fails to reduce the primary deficit by increasing the taxes and reducing the expenditures and transfer payments, Japan would be forced to reduce the value of government debt through either inflation or outright default[76].

In fact, even the rating agencies have been forced to downgrade Japanese government debt in recent years. With the Fitch rating agency downgrading their debt from the highest *AAA* rating in 1998, down to A. And as of 2016, Japan's outlook was dropped to negative – a warning of future downgrades to come.

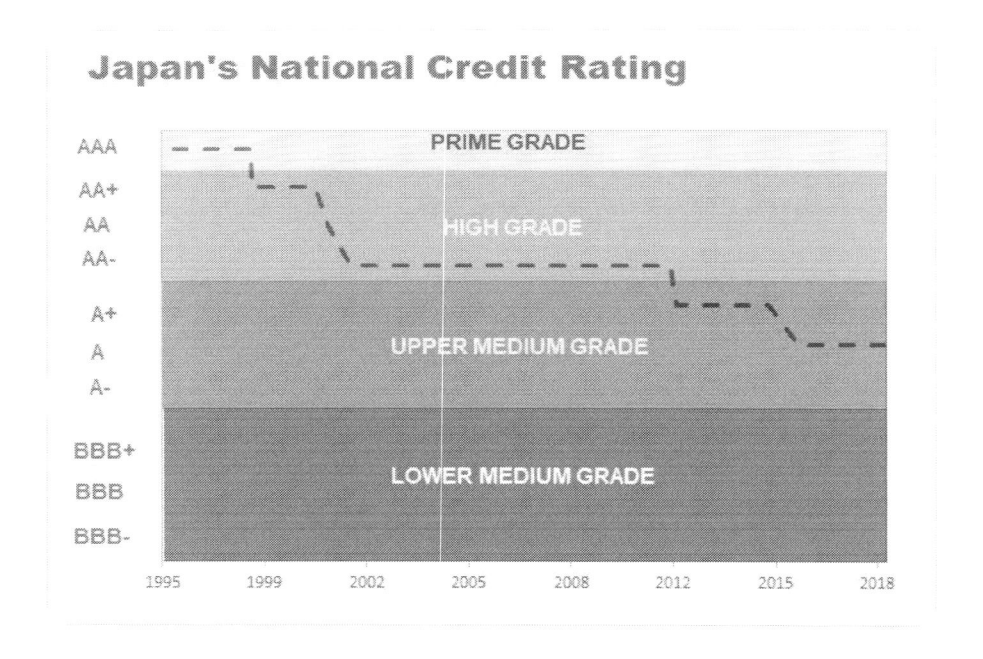

Japan's National Credit Rating

Essentially, there is unanimous agreement that the situation cannot even stabilized without drastic action. There is simply no way to pay off this debt, and once interest rates rise – either from demographic pressures or the BOJ's inability to continue QE because of a lack of eligible JBGs – servicing the debt will require more than the total of Japan's tax revenue.

As far back as 2008, mandatory spending – social security and debt service – exceeded total government revenues. But a significant portion of Japan's budget is taken up just servicing the interest on it's debt. In 2013, Japan spent 25% of its budget (and closer to half of tax revenues) servicing its debt which is currently almost 250% of GDP. This was at an interest rate around half a percent. If their bond yield was to rise even 200 basis points (2%), servicing the debt alone would exceed total government revenues.

Japan's Social Security and Interest Payments vs Revenue

Japan Social Security, Interest Payments, and Revenue as a Percent of GDP,
1965-2027 2018-2027 Projected (Projections Assume 10% Increase in Comsumption Tax - Since Delayed)

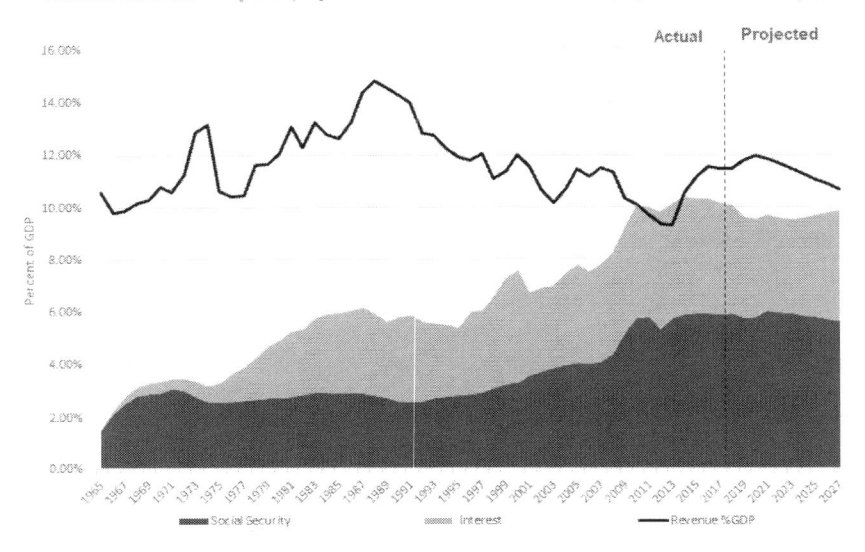

Source: Japanese Ministry of Finance

Tax Revenue Spent on Interest

Percentage of Japanese Total Tax Revenue Spent on Interest Payments on Debt

Source: Japanese Ministry of Finance, Author's Calculations

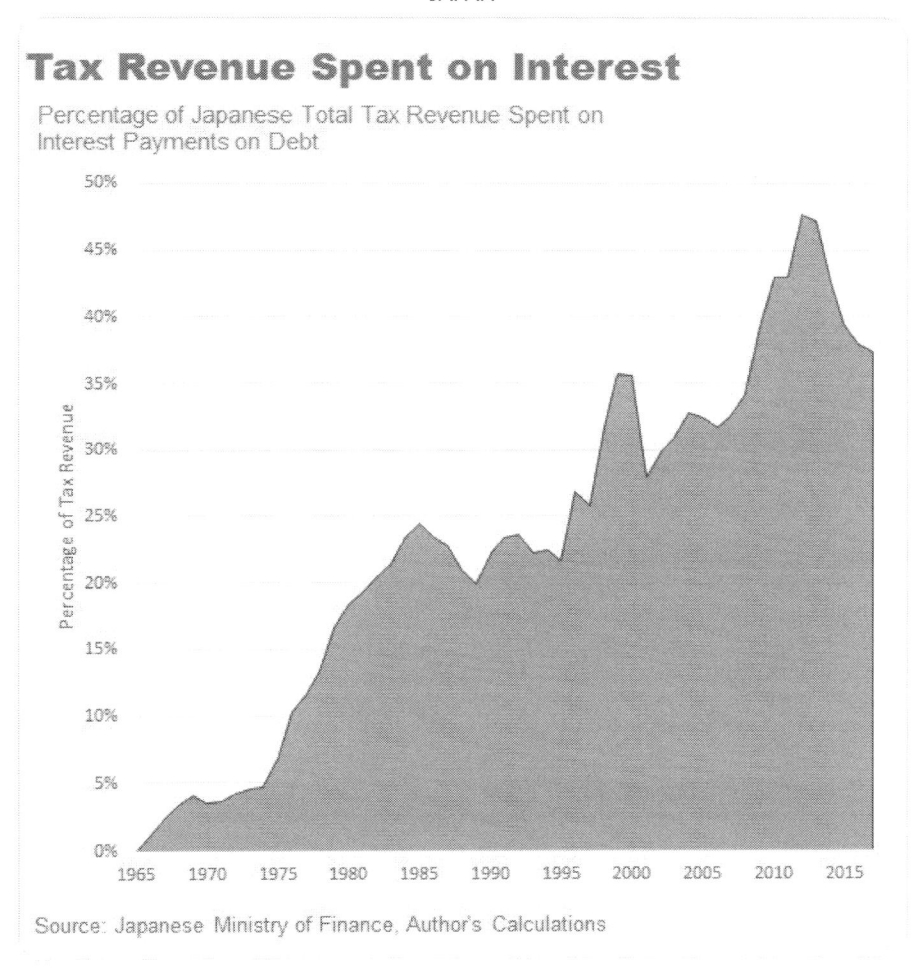

According to Japanese economists, projections show that—unless there is significant fiscal adjustment—the supply of JGBs will outstrip domestic demand for them by the mid-2020s (Hoshi and Ito 2014). Naohiro Yashiro and Akiko Sato Oishi agree, writing in *The Economic Effects of Aging in the United States and Japan*, that:

> *Given the underlying life-cycle framework of the model, popula-*

tion aging will result in a general decline in the national savings rates; the speed of decline will be most noticeable in Japan and will approach zero in the year 2025, when the ratio of the aged to the total population will reach a plateau[77]

The fact that Japan will be facing a wall of demographics is echoed by a 2011 IMF working paper, *Assessing the Risks to the Japanese Government Bond (JGB) Market*, which notes that:

Japan's large pool of domestic savings, stable investor base, and high share of domestic ownership of JGBs have helped maintain stability in the JGB market. **But these favorable factors are likely to diminish over time as population aging reduces household saving and risk appetite recovers. Without a significant policy adjustment, the stock of gross public debt could exceed household financial assets in around 10 years**, *at which point domestic financing may become more difficult. [emphasis added]*[78].

They will soon be in a situation where even if all their debt was forgiven, they would be unable to make even the minimum mandatory expenditures on education and social security without constant infusions of new debt. A situation that will only worsen as their working-age population continues to fall.

JAPAN

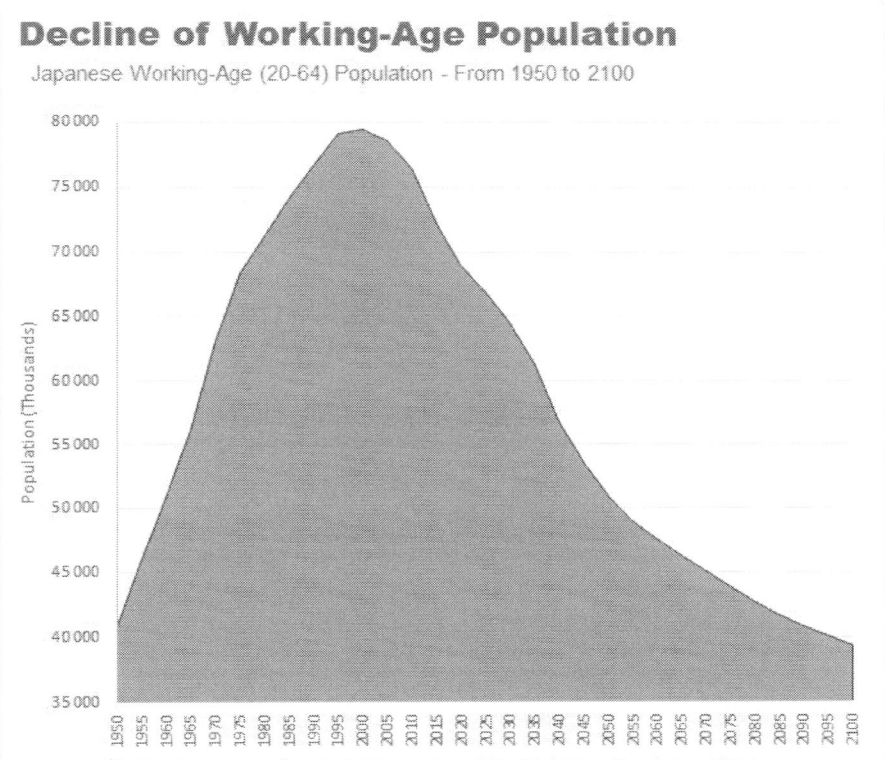

Decline of Working-Age Population

Japanese Working-Age (20-64) Population - From 1950 to 2100

Source: United Nations, Department of Economic and Social Affairs, Population Division (2017). World Population Prospects: The 2017 Revision, DVD Edition.

Japan's Demographic Shift

Japanese Population by Age Group - From 1950 to 2100

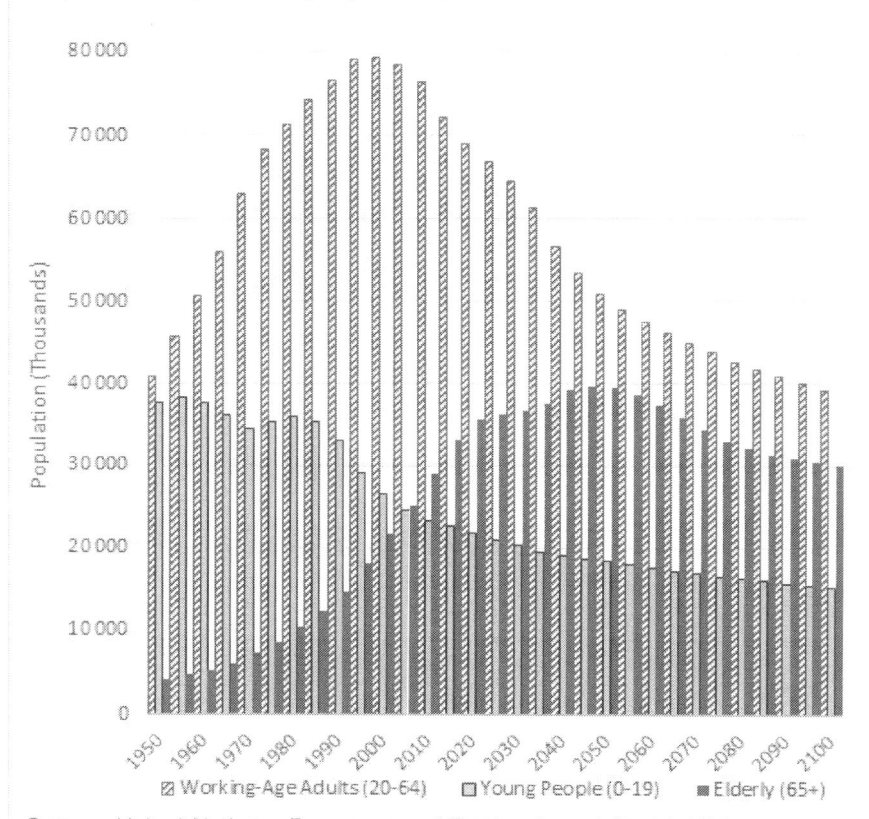

Source: United Nations, Department of Economic and Social Affairs, Population Division (2017). World Population Prospects: The 2017

6

Italy and Europe

The European Union

In 1989 the Berlin Wall fell, and with the collapse of Soviet power, the long dormant *German question* returned. While the United States saw a strong Germany as a counterweight to Russia, The U.K. and France saw things closer to the Soviet viewpoint - that a united Germany was a threat to their national security. Two months before the fall of the Berlin Wall, Margaret Thatcher was in the Soviet Union pleading with Soviet leader Mikhail Gorbachev to intervene and stop the fall of the Berlin Wall and the break-up of East Germany. She told him:

> *We do not want a united Germany. This would lead to a change to postwar borders, and we cannot allow that because such a development would undermine the stability of the whole international situation and could endanger our security*[79],

This was a concern also strongly felt by the French. In a meeting with Thatcher, The French president supposedly said that if East and West Germany were to reunite, Germans would gain more influence than they ever had under Hitler.[80] Despite their protests however, U.S.

president George H.W. Bush supported German reunification – and refused to agree to Soviet hopes that N.A.T.O. would not be extended to the eastern half of the country.

The solution for French and British fears of a newly powerful Germany was the European Union. Tim Marshall, writing in *Prisoners of Geography,* explains that "...the EU was set up so that France and Germany could hug each other so tightly in a loving embrace that neither would be able to get an arm free with which to punch the other.[81]"

In the rush to create the political and monetary union for geopolitical reasons, the planners created a fatal flaw by ignoring the wisdom of economist Nicholas Kaldor, who wrote twenty years earlier in 1971 that:

> *Some day the nations of Europe may be ready to merge their national identities and create a new European Union – the United States of Europe...This will involve the creation of a "full economic and monetary union". But it is a dangerous error to believe that monetary and economic union can precede a political union... For if the creation of a monetary union and Community control over national budgets generates pressures which lead to a breakdown of the whole system it will prevent the development of a political union, not promote it.*[82]"

The introduction of a common currency, without a true political and fiscal union, created perverse incentives throughout the European continent. When the euro was introduced, the bond market came to believe that all EU bonds had the same risk (essentially none). Crucially, EU financial regulators allowed banks to treat Greek debt just like German debt – not even requiring *any* capital reserves against the risk of default until 2006. The graph below shows how the yields from high risk debt such as Greece, Spain, and Italy converged to match France and Germany during this period.

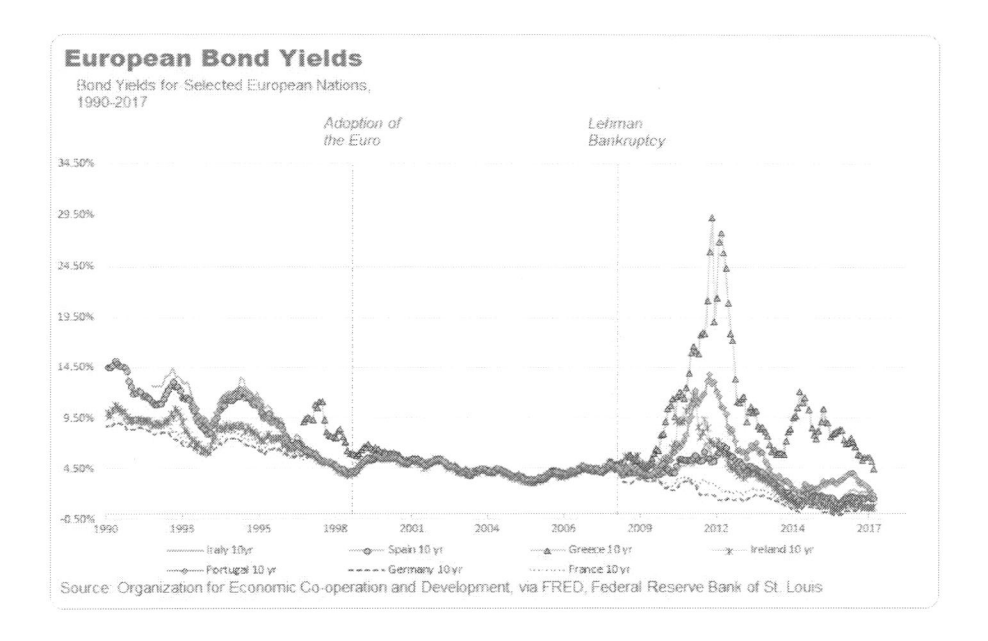

European Bond Yields
Bond Yields for Selected European Nations, 1990-2017

Adoption of the Euro — *Lehman Bankruptcy*

Italy 10yr, Spain 10 yr, Greece 10 yr, Ireland 10 yr, Portugal 10 yr, Germany 10 yr, France 10 yr

Source: Organization for Economic Co-operation and Development, via FRED, Federal Reserve Bank of St. Louis

This allowed nations to borrow increasing amounts of euros at incredibly low, and totally unwarranted interest rates. It wasn't until after the Great Recession that the error of this was realized - at which point these nations had spent a decade borrowing vast sums.

The other unique problem for heavily indebted European countries is that they do not control the currency that their debt is issued in. Unlike the U.S. or Japan, nations like Spain, Italy, and Greece cannot simply accept higher inflation and print currency to pay their debts. The Euro is out of their direct control. When Europe created the euro it critically left countries independent to set their own fiscal policies and to issue their own debt. While the single currency union promoted trade between the countries of Europe, it also created a situation where countries would be unable to lower their debt burden simply by devaluing their currency. Almost uniquely in the post-Bretton Woods world, the individual EU nations do not control the currency that they issue bonds in.

To put the scale of the problem the EU faces into perspective, value of the entire American subprime mortgage market was estimated at $1.3 trillion as of March 2007 (1.57 trillion in 2018 dollars), The debt of the countries known collectively as the PIIGS—Portugal, Ireland, Italy, Greece, and Spain—whose debt was put into question during their 2010 sovereign debt crisis—totals 4.85 trillion as of 2018. And of course, the United State's annual GDP is 18.62 trillion while the combined PIIGS GDP is only 3.76 trillion. The St. Louis Federal Reserve highlighted the exposure to PIIGS debt by various European countries and the U.S.:

> *To get a better sense of the risks, economists often express these amounts as a percent of the creditor country's GDP. By this metric, French banks had the most exposure (32 percent), followed by Dutch banks (26 percent) and then German banks (20 percent). The exposure of U.K. banks was 17 percent, and the exposure of U.S. banks was only 1 percent*[83].

This debt not only threatens the economies of the European nations, but the political union itself. Southern Europe will be increasingly incentivized to exit the union, leave the euro, and monetize their debt.There is an additional political complexion to the crisis. Germany has low levels of debt and a large trading surplus. And large amounts of debt from Southern Europe are held by German banks. If the southern European nations monetized their debts Germany would pay the price and would effectively be subsidizing the structural problems of their economies.

But the interesting part of the story is the mechanics of exactly these structural problems in Southern Europe arose. While the regulatory failure which brought bond yields down for Southern Europe *allowed* these nations to accumulate massive debts, it was *incentivized* not by the emergence of massive trade imbalances within the continent.

When the Euro was created, Germany was determined to bring down the high levels of unemployment that it saw after the reunification with East Germany. The red-green government in Germany put political pressure on labor unions to restrict the growth of nominal and real wages in order to boost employment. As Heiner Flassbeck, the former chief macroeconomist at the German Institute for Economic Research, wrote in piece in *American Affairs:*

> *As the unions bent to political pressure, the growth of nominal wages was halved compared to the decade before. This brought about huge divergences in nominal ULC [Unit Labor Costs] developments among the members of the EMU [European Monetary Union]...Even though the annual divergences in national increases to ULCs were relatively small, over time, these "small" annual differences accumulated to produce dramatically large gaps between countries. At the end of the first decade of the EMU, the cost and price gap between Germany and southern Europe amounted to some 25 percent, and the gap between Germany and France was 15 percent. In other words, Germany's real exchange rate had depreciated significantly, even though national currencies no longer existed within the EMU*[84].

Increase in Unit Labor Costs

Country	Increase
Greece	40%
Spain	31%
Portugal	27%
Germany	7%

Source: German Institute for Economic Research

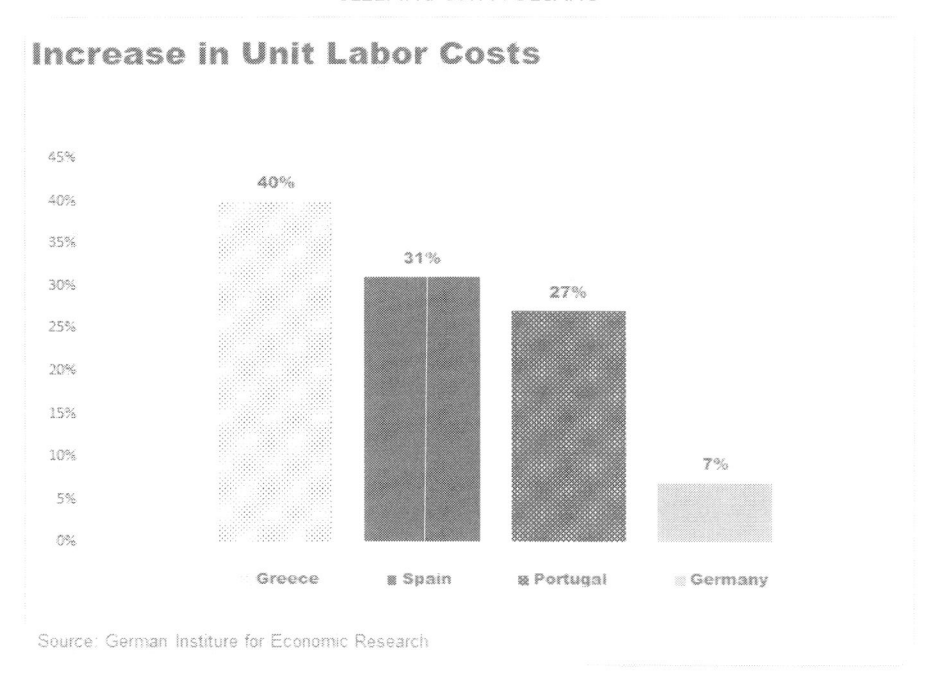

In Europe, the German's high savings rate can be explained by the national polices which held down wages relative to other countries. Lowering German wages functions to reduce German household consumption and subsidizes export industries – which serves to increase the German savings rates and thus boost their exports.

All of this came to in head with the Great Financial Crisis. The 2008 financial crisis damaged the economies and budgets of nations around the world – many nations were forced to take on massive amounts of debt to bailout their financial systems. But in Southern Europe, the recession and credit crunch was enough to push governments themselves into insolvency. By 2010 the Southern European nations were in the midst of a sovereign debt crisis. Portugal, Italy, Ireland, Greece, and Spain, (known in financial circles as the PIIGS), saw the value of their debt plummet and their interest rates to borrow soar as their financial status deteriorated.

PIIGS Bond Yields After Soverign Debt Crisis

PIIGS (Portugal, Italy, Ireland, Greece, and Spain) 10-year Bond Yields. Before and after 2010 Soverign Debt Crisis

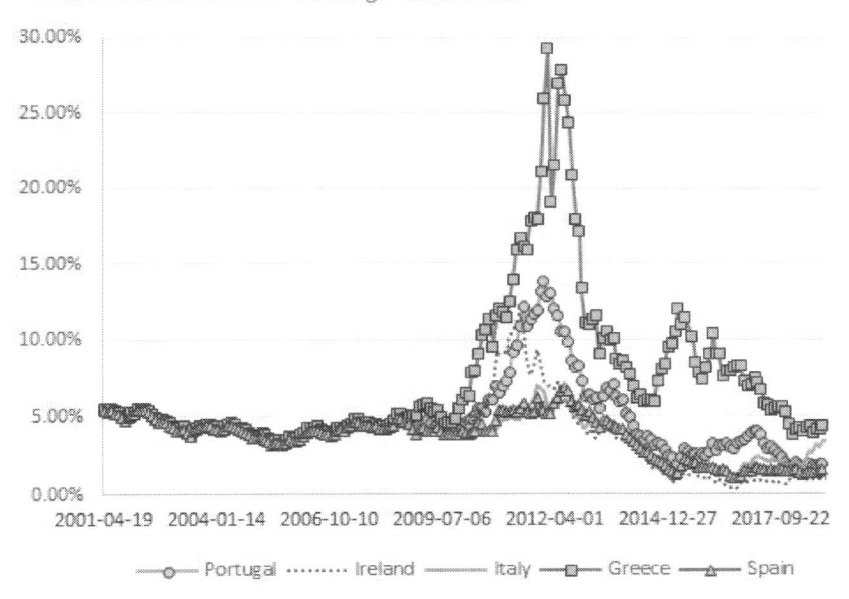

Source: Organization for Economic Co-operation and Development, via FRED, Federal Reserve Bank of St. Louis

The market realized that there was a substantial risk they could not pay their debts, and so, quite suddenly, the interest they demanded to lend to these countries rose dramatically. This crisis has not been resolved. In fact, today Greece's debt to GDP ratio is significantly higher than what it was *before* the crisis.

Greece's Increasing Levels of Debt

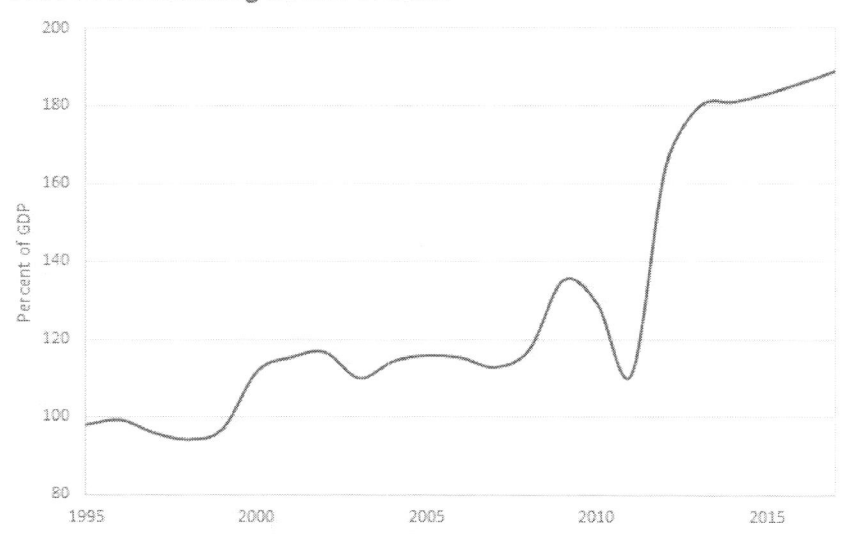

The situation is even less sustainable than before – but the reckoning has been delayed. The European Central Bank (ECB) purchased the bonds of Greece and the other troubled nations and pushed their yields down artificially with QE. To the point that in May 2019 the yield on US treasury five year bonds are higher than on Greek five year bonds. Not because Greek debt has less risk of default than US debt, but because the yield has been artificially manipulated lower. The ECB has continued to purchase the bonds of European countries –propping up their value and preventing a systematic collapse of the European banking system. Mervyn King, the Bank of England Governor at the time of the crisis, explained that:

> *Dealing with a banking crisis was difficult enough, but at least there were public-sector balance sheets on to which the problems*

could be moved. Once you move into sovereign debt, there is no answer; there's no backstop[85].

The crisis began when, In 2009, Greece projected its budget deficit would be 12.9% of their GDP - Leading ratings agencies to downgrade their bonds and sparking a crisis. For Greece, the interest rate they needed to pay to borrow for a 10-year note jumped from less 5% to almost 30%. But, this was not just a crisis for Greece. As John Maynard Keynes said, "If you owe your bank a hundred pounds, you have a problem. But if you owe a million, it has." Greece owed billions of euros to French and German Banks. If Greece could not pay, not only Greece, but the French and German Banking systems would be insolvent.

Former Greek finance minister Yanis Varoufakis has denounced rescue plan for Greece as a modern version of the Versailles Treaty. In his 2017 book, *Adults in the Room*, detailing his time as Greek finance minister at the height of the crisis, he wrote:

> *Greece's endemic underdevelopment, mismanagement, and corruption, explain its permanent economic weakness. But its recent insolvency is due to the fundamental design faults of the EU and its monetary union the euro...While the drachma devalued, these deficits were kept in check. But when it was replaced by the euro, loans from German and French banks propelled Greek deficits into the stratosphere.*

> *The Credit Crunch of 2008 that followed Wall Street's collapse bankrupted Europe's bankers who ceased all lending by 2009. Unable to roll over its debts, Greece fell into its insolvency hole later that year. Suddenly three French banks faced losses from peripheral debt at least twice the size of the French Economy...For every thirty euros they were exposed to, the had access to only one...The same three French bank's loans to the Italian, Spanish*

and Portuguese governments alone came to 34 percent of France's total economy...If the Greek government could not meet its repayments, money men around the globe would get spooked and stop lending to the Portuguese, possibly to the Italian and Spanish states as well, fearing that they would be the next to go into arrears. Unable to refinance their combined debt of nearly €1.76 trillion at affordable interest rates, the Italian, Spanish and Portuguese governments would be hard pressed to service their loans to France's top three banks...[86]

German banks had the same issues as the French banks. The crisis in Southern Europe threatened to destroy the economies of every nation in Europe. The solution, was to ensure that Greece did not default and cause a domino effect throughout Southern Europe. The EU provided a *rescue fund* for Greece which essentially amounted to providing them a loan with which they could service their existing loans to their French and German creditors. But there was no restructuring of the debt owed to the foreign banks.

This so called *bailout*, and the austerity measures that went with it were put to the Greek people in a referendum on July 5th 2015, which was promptly defeated at the polls. However, the EU essentially threatened to shut down the Greek banking system if Greece did not accede. ATMs would not provide cash, depositors would be unable to access their bank accounts, and the Greek economy would collapse overnight. The Wall Street Journal reported days after the referendum, on July 12th 2015:

Europe's ultimatum to Greece, demanding full capitulation as the price of any new bailout, marks the failure of a rebellion by a small, debt-ridden country against its lenders' austerity policies, after Germany flexed its muscles and offered Athens a choice between obeisance or destruction.

Sunday's statement on Greece by eurozone finance ministers will go down as one of the most brutal diplomatic démarches in the history of the European Union, a bloc built to foster peace and harmony that is now publicly threatening one of its own with ruination unless it surrenders[87].

While the measures have allowed the European system to maintain the fiction that Greece and the rest of Southern Europe can repay their debts, the impact on Greece has been nothing short of horrific. Under the austerity plan, the country has not managed to substantially reduce its debt. In fact Greece's debt to GDP ratio has jump from 112% in 2007 to over 188% by 2017. Loans have been extended not restructured or forgiven. The nation was forced by the EU to enact twelve rounds of tax increases, spending cuts, and reforms - sparking riots and nationwide protests. In many ways the situation in Greece has surpassed the Great Depression in the United States. With Greek unemployment spiking higher and lasting longer than in the United States during the 1930s' Great Depression.

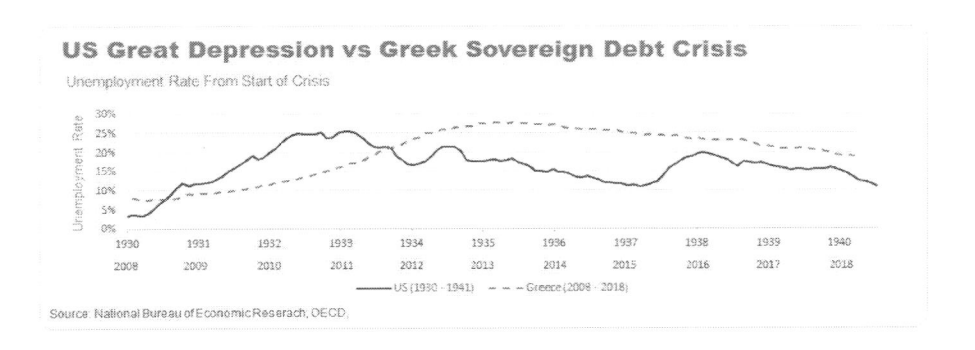

Greece's minimum wage was cut by twenty-two percent. Pensions have been reduced forty percent. The national healthcare budget was slashed by roughly forty percent as well. Suicides have more than doubled since the crises began. Even Infant mortality rates which had been falling for decades have increased thirty-four percent in recent years. Despite all this, Greece has not moved to a sustainable fiscal position, only managed to continue to service its debt to its creditors.

* * *

The fate of Greece stands as a warning to the world. Failure to stabilize debt can have horrific consequences, and losing the confidence of the market can happen in a matter of months.

Greece of course is a small nation which had the worst debt to GDP ratio in Europe before the crisis. The world has much more confidence in larger, less corrupt, nations with healthier financial status. And crucially, Greece did not control the currency that its debt was issued in. Greece was forced into austerity because more powerful nations, who controlled the Greek banking system, insisted on it. Few other nations could be so bullied. It is unlikely, almost politically unthinkable, that most nations would institute harsh austerity to pay down their debts and support creditors. But that does not mean that their debts are sustainable, only that they have options Greece did not have.

One nation in Europe in particular looks to be the focus of any future crisis is Italy. Italy is a nation that is *too big to fail.*

The Italian Crisis

That is the background to the situation today. The sovereign debt crisis that began in 2010 remains just under the surface. Since 2010 the ECB has used unconventional monetary policy to drive long-term bond yields to ultra low rates. This has allowed Greece, and Italy and the rest of the nations in Southern Europe to pay the interest on their debts. But even with these low interest rates, Debt to GDP has increased substantially.

And combined with these low interest rates have been harsh austerity measures. This austerity has stunted growth for almost a decade throughout Southern Europe. For the last twenty years Italy's average annual growth rate has been zero[88]. Italy's unemployment rate has been over ten percent for the last seven years -with youth unemployment over thirty percent. Its hardly surprising that Italy has voted in a populist government opposed to austerity.

The nation is currently governed by a coalition of left-wing and right-wing populists who's 2019 budget plans to cut taxes and further lower the retirement age. According to a Reuters report on the Italian budget:

The new budget will thus set aside 10 billion euros for a so-called citizens' income, a handout for the poor championed by the anti-establishment 5-Star Movement. The spending plan will also allow ageing Italians to retire early and offer tax cuts to about one million self-employed workers, in deference to the hard-rightist League.

Their budget plans have brought the populist parties into conflict with the rest of the EU over their budget. And now they are proposing they can pay for their plans by issuing their own parallel currency. The New York Times reported recently that:

...one proposal has caused particular consternation and raised fresh concerns that Italy, the third largest economy in the eurozone, could blow up the entire bloc. That land mine, critics say, is called the mini-BOT[89]

In Italy the proposal to create a parallel currency, called mini-Bots, looks like it might become a reality. This idea is a threat to the existence of the eurozone – and the stability of the entire global economy.

While this idea is not a new one, it has recently become a serious proposal[90] in Italy as populist Euro-skeptic parties have gained more and more control. Salvini, Italy's deputy prime minister and leader of the League, might win the next election with such a margin that the rules will give him a super-majority in Parliament. Allowing him the create the parallel currency that they've been talking about for years. Reuters Reported back in 2017 that:

> *The Northern League's Borghi said Italy "has to be ready for the euro's collapse," which he sees as only a matter of time. He is the architect of the party's proposal - which Berlusconi has also hinted he would support - called "mini-BOTs", named after Italy's short-term Treasury bills.*
>
> *Borghi says initially some 70 billion euros of these small denomination, interest-free bonds would be issued by the Treasury to firms and individuals owed money by the state as payment for services or as tax rebates. They could then be used as money to pay taxes and buy any services or goods provided by the state, including, for example, petrol at stations run by state-controlled oil company ENI*[91].

Issuing a currency in parallel to the Euro risks the breaking up the European monetary union - if only because having a parallel system in place would remove much of the pain from a transition. Europe's control over the euro, and threats to use that power to shut down Greece's banking system, is what forced the Greek populists to submit to austerity back in 2013. The Italians have obviously taken this lesson to heart - which is why the proposed parallel currency is causing such consternation in Europe.

Northern League's economics spokesman Claudio Borghi told reuters:

With a parallel currency in place, if we want to leave the euro our economy will still be able to operate even if the European Central Bank tries to crush us by shutting off liquidity in euros[92]

In a recent Bloomberg Opinion Piece, a former member of the Financial Times editorial staff, Ferdinando Giugliano wrote:

> *The fear is that an Italian euro departure might be prepared secretly and announced abruptly. One would usually expect a government to hold a referendum on such a big decision, as Britain did on its EU membership, but Italy's populists might be wary of that approach because of concerns that even announcing a vote may lead to bank runs and capital flight (which would no doubt scupper any "leave" campaign).*
>
> *This is why the idea of mini-BOTs captures so much attention*[93].

Italian Fundamentals

Southern European nations including Italy, are much more constrained than the United States. Their creditors, nations such as Germany, control the printing of euros and are unwilling to allow these countries to de facto default on their obligations by increasing the money supply. Italy is certainly not the nation with the worst financial outlook in the eurozone – but its size, the third largest economy in the E.U., makes it much more of a risk to the system than other troubled nations. There is no way to bail out a country of Italy's size. Furthermore, according to economists at the Peterson Institute for International Economics:

> *...none of the powerful stabilization instruments that the euro area has developed over the years could be deployed to rescue Italy. Following crisis-related downgrades, Italy would no longer be eligible for the ECB's quantitative easing bond-purchasing pro-*

gram. The ECB would stop accepting Italian bonds as collateral. Access to emergency support programs—the ESM, and through it, the Outright Monetary Transactions (OMT) program—would be conditional on fiscal adjustment, the opposite of what Italy's new government has promised. Unless the government were to change course, it would be forced to exit the euro, even if this is not its current plan. And second, there is Italy's interconnectedness and size. With the ECB using all available tools to limit contagion, the euro might survive Italexit. But an exit would nonetheless put Italy, the euro area economy, and the European Union in deep distress.

Italy has a poor financial outlook in large part because for decades Italian spending has surpassed revenues; with yearly deficits of more than four percent of GDP not uncommon. Spending on social programs and a bloated public sector workforce has driven deficits for decades.

Italy's Spending and Revenue

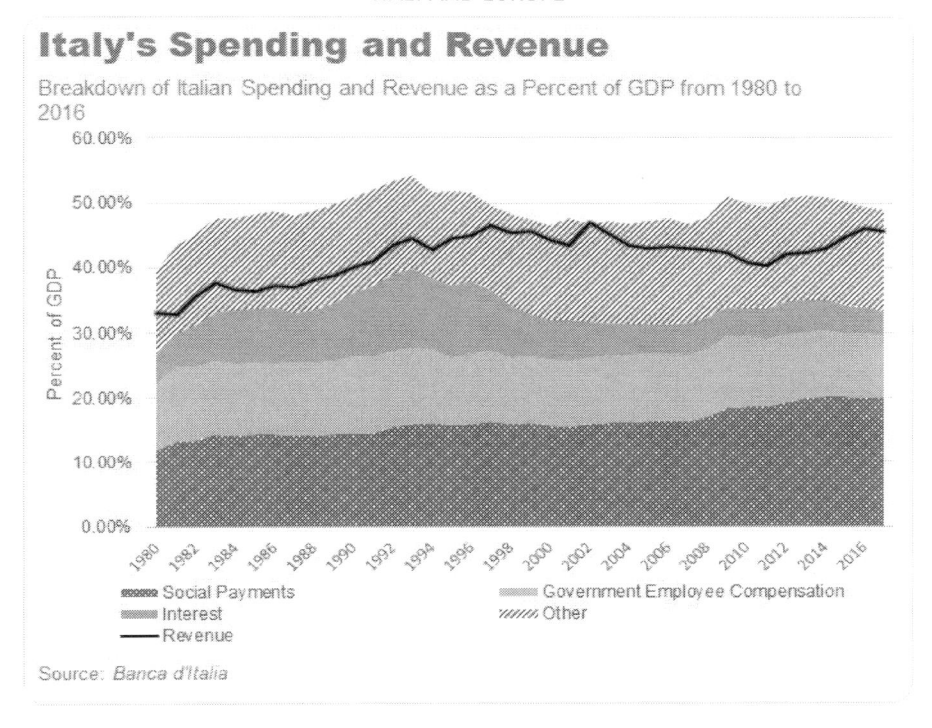

Breakdown of Italian Spending and Revenue as a Percent of GDP from 1980 to 2016

Legend: Social Payments, Interest, Revenue, Government Employee Compensation, Other

Source: *Banca d'Italia*

Over the decades these persistent deficits have steadily increased the Italian Debt to GDP ratio. The Italian fiscal picture has only stabilized recently because the ECB has been buying Italian bonds and forcing the the interest rate on the debt lower with its quantitative easing program. In fact, since the 2010 sovereign debt crisis began, the ECB has been the only *net* buyer of Italian Debt.

Italy's Government Debt

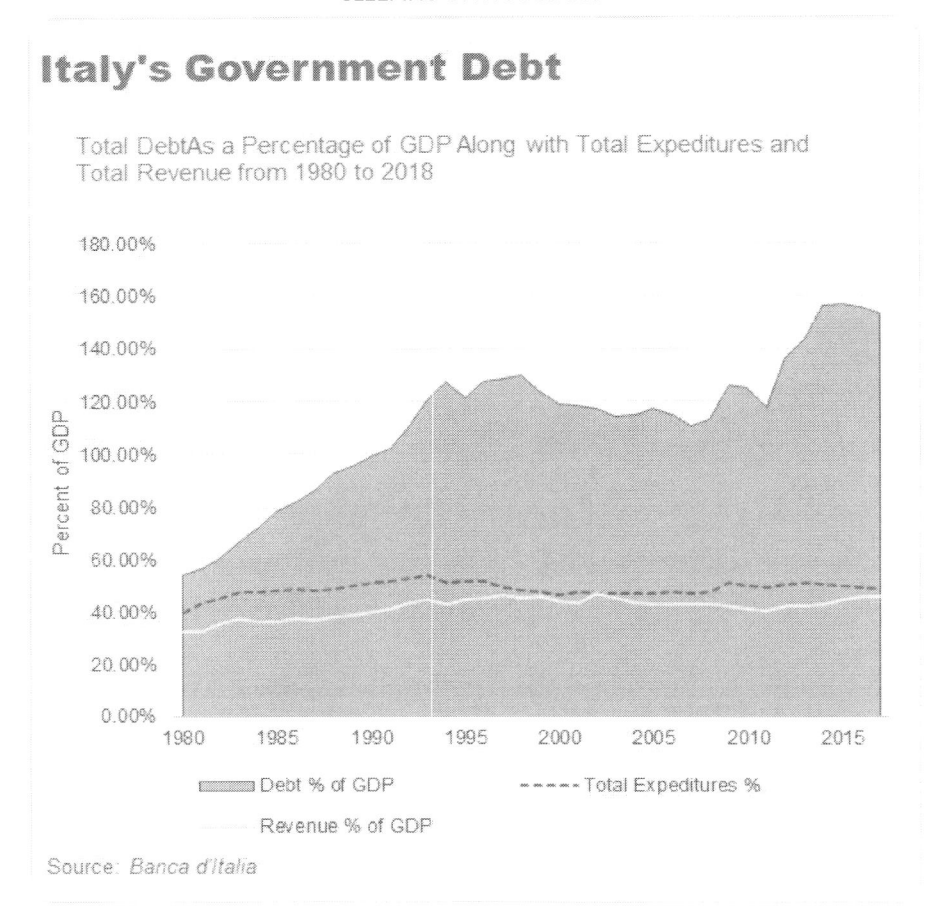

Total Debt As a Percentage of GDP Along with Total Expeditures and Total Revenue from 1980 to 2018

Source: *Banca d'Italia*

The crucial problems for the Italians is that the Italian demographic picture leaves no room to hope that they will be able to grow themselves out of the crisis. They will instead see their costs for social programs soar as their population ages. And their plummeting working-age populations means that tax revenues will continue to fall for the foreseeable future.

Italy's Working-Age Population

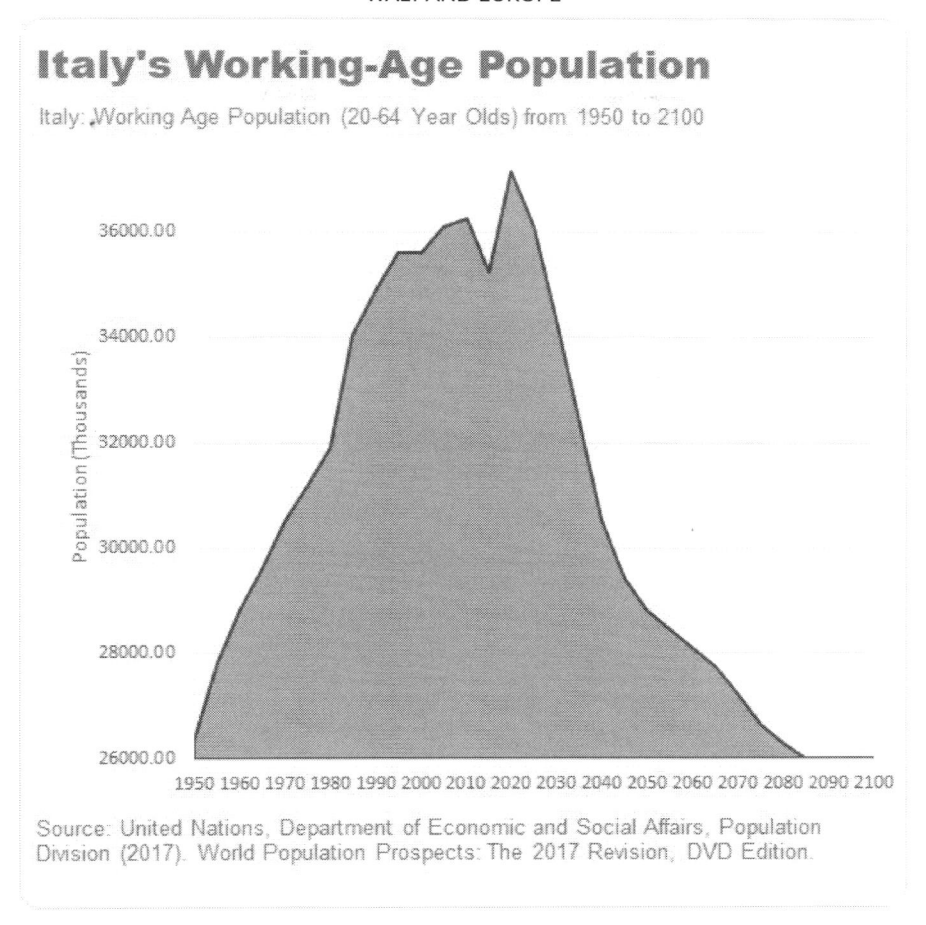

Italy: Working Age Population (20-64 Year Olds) from 1950 to 2100

Source: United Nations, Department of Economic and Social Affairs, Population Division (2017). World Population Prospects: The 2017 Revision, DVD Edition.

Despite the bleak fiscal future, and the growing economic risks, Italy has taken no moves to reform it's structural economic problems. Italy has been swept by populism and is currently governed by a coalition of left-wing and right-wing populists who's 2019 budget plans to cut taxes and further lower the retirement age. According to a Reuters report on the Italian budget:

> *The new budget will thus set aside 10 billion euros for a so-called citizens' income, a handout for the poor championed by the anti-*

establishment 5-Star Movement. The spending plan will also allow ageing Italians to retire early and offer tax cuts to about one million self-employed workers, in deference to the hard-rightist League.

The IMF's April 2019 Global Financial Stability report highlighted the Italian banking system increasing dependence on sovereign debt.

Financial channels between sovereigns and banks have strengthened in countries with more indebted sovereigns. Domestic government bond portfolios of banking systems are large relative to assets in several countries, particularly Belgium, Italy, Portugal, and Spain ...This may partly reflect the higher yields on government bonds in many of these countries, the use of these bonds as collateral for central bank liquidity facilities, zero risk weights on sovereign bonds (which enable government bond portfolios to increase without reducing Tier 1 capital ratios), and liquidity regulations (which treat government bonds as liquid assets). Data for the banks included in the European Banking Authority's Transparency Exercise (EBA banks) also reveal that the proportion of lower-rated government bonds held by Italian and Portuguese banks, in particular, has increased following downgrades to sovereign credit ratings... The rising exposure to government bonds, and downgrades to sovereign credit ratings, have made banks in some countries more vulnerable to sovereign shocks

The Threat to Europe and the Rest Of the World

Former IMF deputy director Desmond Lachman recently laid out the unenviable situation that Italy finds itself in today:

Italy's basic policy dilemma boils down to a choice between two unattractive options. Should Italy try to stimulate its economy through an expansionary fiscal policy even though that might give fuel to the country's bond vigilantes who are already concerned about Italy's budget deficit and its very high public debt-to-GDP ratio?

Or should Italy again try to address its shaky public finances with budget austerity even though that risks deepening any economic recession? Such a deepening in turn might raise serious questions about the country's ability to service its public debt mountain.

...All of this suggests that Italy could be heading soon for a sovereign debt crisis that could have serious implications for the global economy. Being 10 times the size of the Greek economy and having the world's third-largest sovereign debt market, Italy has the real potential to trigger a global financial market crisis[94].

In an interview for this book, Lachman expanded on his concerns for Italy:

This is really kind of quite a big deal... Italy is ten times the size of the Greek economy. And it has ten times the amount of debt. We saw when there was a Greek debt crisis back in 2010, that it really had big implications for the global economy. People got really very unsettled. Markets were very stressed. But Italy is ten times the size of Greece, so if Italy were to default on its debts that would be a major event.

Another way you can look at it is that it Italy's sovereign bond market is the third largest in the world – after the United States and Japan. So there's like two and a half trillion dollars of debt that the Italian government has. And something like a trillion dollars of that is held by foreign banks and foreign investors; so if there is a default on that you'd get a banking crisis in Europe.

So what I'm saying is that in my view this is an event – if they were to default –that would be an event big enough to trigger a global economic downturn.

Lachman also raised concerns about Itay's current proposal to issue a type of parallel currency to the euro, a move that would be perceived as a major step towards leaving the euro. He said:

From the political point of view you've got a populist government that doesn't care that they are on a collision course with the Europeans. They're talking about issuing – they're calling them mini-BOTs; it's like a parallel currency. You're getting a lot of red flags that things in Italy are going in the wrong direction.

If Europe broke up that would be like a nuclear bomb going off in the middle of Europe – it would really be a disaster. All of the European banks would be – there would be major major stress. If the euro were to break up Italy would certainly default on its debt. because the interest rates in Italy would just shoot up. If that were the case you'd have a lot of banks that are holding Italian paper – a lot of the French banks are holding Italian paper – they'd be in deep trouble. It would be an event like Lehman and blowing up.

Looking at the demographics for Europe, and the size of the debt burdens in the continent, it seems unlikely that Europe will be able to stabilize its debt in the coming decades. It remains to be seen if this parallel currency issue will be the factor that brings down the monetary union, but in the long run it seems inevitable that the monetary union cannot survive.

7

China

The Rise of China

After the communists came to power in China, Mao collectivized Chinese agriculture – outlawing the private ownership of farms. Mao also dictated counterproductive polices like *deep plowing* which was harmful in shallow soil, and the *great sparrow campaign* which sought to eliminate birds as pests but resulted in insect numbers skyrocketing.

During this *Great Leap Forward,* as the process was called, forty-five million people in China would ultimately starve to death during this period. Xiaogang, a small village in the East Anhui Province which had a population of 120 before 1958, saw 67 villagers die of hunger. In the face of this ongoing starvation, eighteen farmers in the small Xiaogang village signed a secret agreement, which if discovered would likely mean their deaths, to abandon the communist collectivization and divide communally owned farmland into individual pieces of private property called household contracts. The leader of this group, a farmer named Yan, recounts their decision to divide the land into private plots:

Villagers tended collective fields in exchange for 'work points' that could be redeemed for food. But we had no strength and enthusiasm to work in collective fields due to hunger. We even didn't have time because we were always being organized by governmental work teams who taught us politics.

I was selected as the deputy leader and later as the leader of our production team in 1962. The grain output in our village was 15,000 kilograms per year before launching the household contract system. Food was not adequate to feed everyone. Families boiled tree leaves, bark and any edible wild plants; we ate whatever we could find. After consulting with some other villagers, I made up my mind to contract land to individual households no matter what penalty would be imposed on me. We didn't want to starve anymore[95].

The abandonment of collectivism was wildly successful. Under the collectivized system, the farmers had no incentive to work beyond the minimum amount required. With the return to private property effort was rewarded. Grain output increased over six times over the previous year. The per capita income of Xiaogang climbed to 400 yuan from 22 yuan. When the villager's scheme came to light, they were accused by some of "digging up the cornerstone of socialism" but with the death of Mao, China was ready for reform. Xiaogang was permitted to continue and soon privatization spread to other farms and to the entire nation.

Ultimately Mao's successor Deng would implement substantial market-based reforms throughout China, which when combined with the incredibly sharp fall in fertility rates and currency manipulation would see China's share of GDP soar to incredible heights with amazing speed. In terms of purchasing power parity (PPP), which attempts to remove currency exchange rates from the equation and reflects purchasing power for a basket of goods, China already has the largest share of world GDP.

Share of World GDP
(PPP) Selected Countries From 1950 to 2023

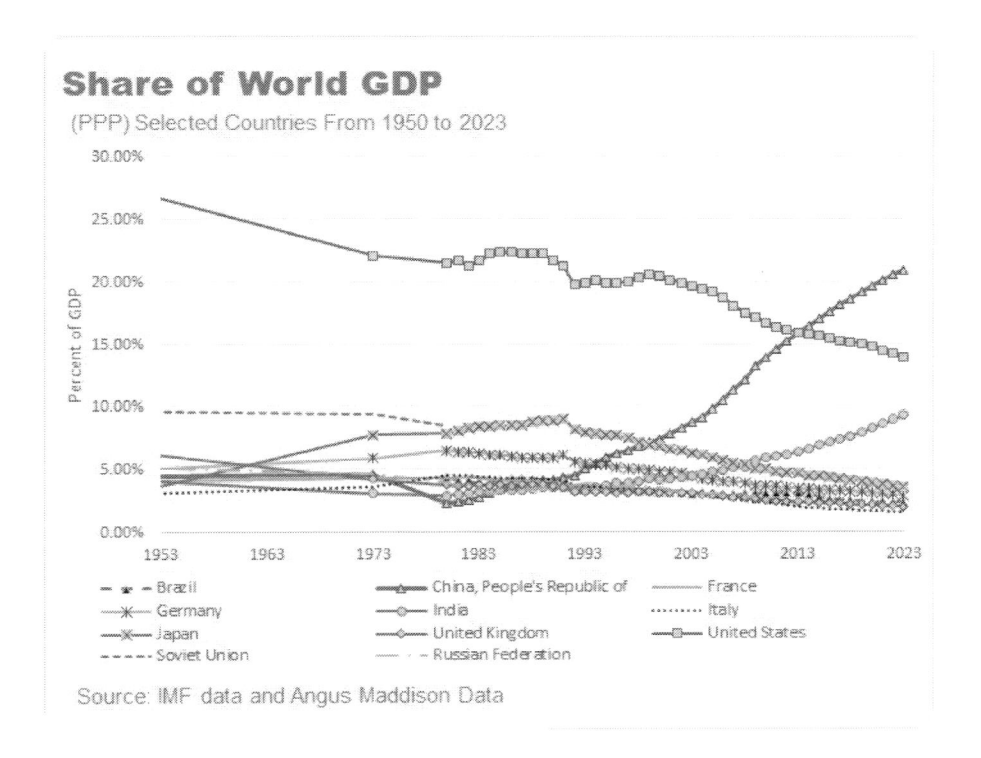

Source: IMF data and Angus Maddison Data

While market reforms – combined with demographics – has allowed the Chinese economy to thrive, China however has never fully embraced capitalism. Nor have they opened up their economy to the rest of the world. A 2012 report from the Economic Policy Institute highlights the failure of China to open their market to the United States following entry into the WTO. The report states in part that:

> *China's entry into the WTO in 2001 was supposed to bring it into compliance with an enforceable, rules-based regime that would require China to open its markets to imports from the United*

States and other nations by reducing tariffs and addressing non-tariff barriers to trade. Promoters of liberalized U.S.-China trade argued that the United States would benefit because of increased exports to a large and growing consumer market in China.... However, as a result of China's currency manipulation and other trade-distorting practices, including extensive subsidies, legal and illegal barriers to imports, dumping, and suppression of wages and labor rights, the envisioned flow of U.S. exports to China did not occur. Further, the agreement spurred foreign direct investment in Chinese enterprises, which has expanded China's manufacturing sector at the expense of the United States. Finally, the core of the agreement failed to include any protections to maintain or improve labor or environmental standards or to prohibit currency manipulation.

While many countries today can fairly be accused of exchange rate manipulation, no nation is more obvious an offender than China. China's herculean efforts to hold down the value of the yuan have resulted in the biggest build-up of currency reserves in history. With china now holding an amount well in excess of three trillion dollars[96].

Debt Binge in China

China has a lot of problems these days. A trade war with their largest trading partner, worldwide backlash against their *Made in China 2025* plan, the rapid aging of their population, and worrying signs that the Chinese economy is slowing down.

How exactly these issues will play out for the Chinese economy is difficult to predict, in large part because of the lack of reliable data. Even the basic fact of China's GDP is suspect. A recent Bookings Institute study found that China's economy is likely 12% smaller than official Chinese data suggests[97]. Considering how this failure to report reliable data impacts everything from from debt to GDP ratios to import and export numbers, getting an accurate read of the Chinese economy is extremely difficult.

But even with their manipulated data, we can see the China is about to begin rapidly aging.

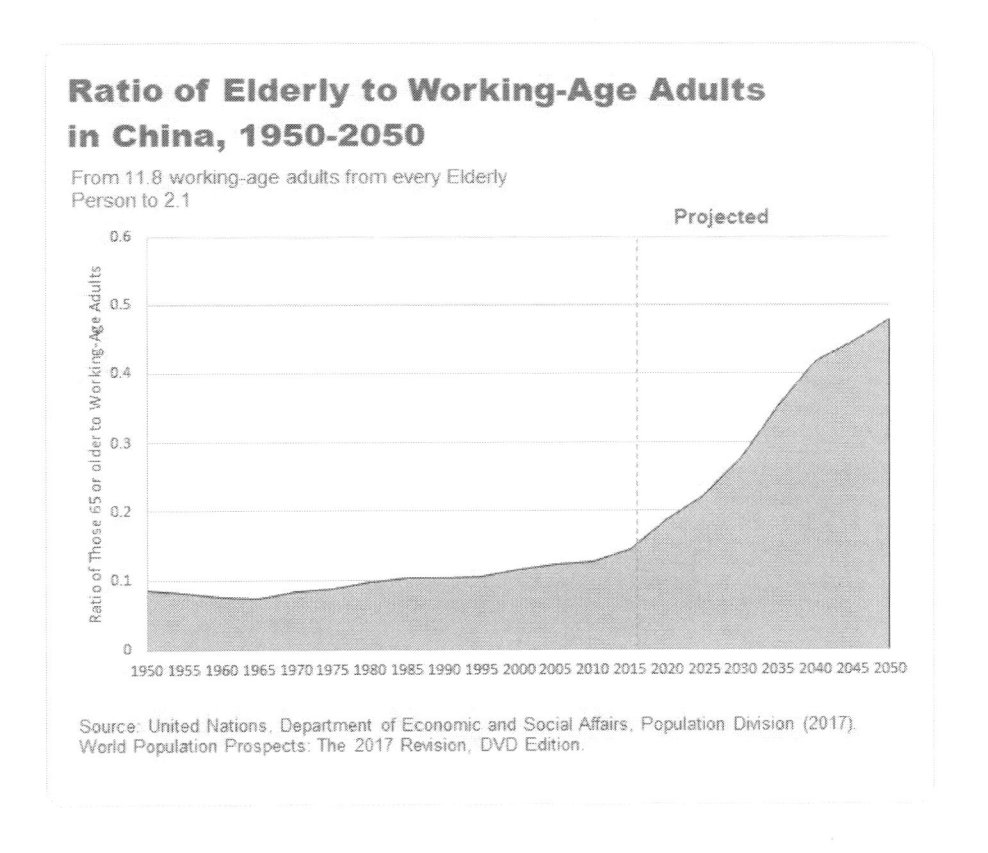

Ratio of Elderly to Working-Age Adults in China, 1950-2050

From 11.8 working-age adults from every Elderly Person to 2.1

Source: United Nations, Department of Economic and Social Affairs, Population Division (2017). World Population Prospects: The 2017 Revision, DVD Edition.

In a recent article in Foreign Affairs, *With Great Demographics Comes Great Power*, Nicholas Eberstadt writes:

> *As China's working population slumps, its over-65 population is*

set to explode. Between 2015 and 2040, the number of Chinese over the age of 65 is projected to rise from about 135 million to 325 million or more. By 2040, China could have twice as many elderly people as children under the age of 15, and the median age of China's population could rise to 48, up from 37 in 2015 and less than 25 in 1990. No country has ever gone gray at a faster pace. The process will be particularly extreme in rural China, as young Chinese migrate to the cities in search of opportunity. On the whole, China's elderly in 2040 will be both poor and poorly educated, dependent on others for the overwhelming majority of their consumption and other needs[98].

On top of their demographic challenges, to understand China's position, you have to understand how, since the Great Recession, China has been accumulating massive amounts of debt. Jamil Anderlini, writing in the Financial Times, explains that:

In the aftermath of the global financial crisis, China's manufacturing and export dependent economy crumbled and the ruling Communist Party panicked. Party leaders estimated they needed to sustain a minimum annual growth rate of 8 per cent if they were to contain political unrest that could threaten authoritarian rule. The solution was to unleash what economists have called the greatest example of monetary easing in history — an enormous wave of easy loans channeled through the state-owned banking system[99]

Indeed, since 2008, Chinese debt has exploded from 7.7 trillion to 33 trillion today (adjusted for inflation) – and an almost doubling of their debt as a percent of GDP.

China's Debt Accumulation

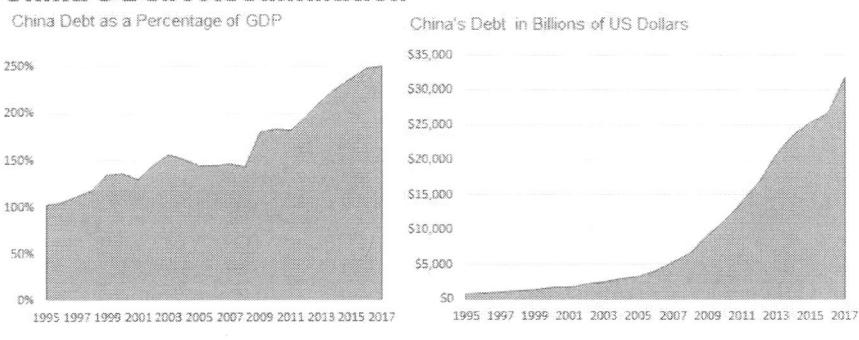

China Debt as a Percentage of GDP

China's Debt in Billions of US Dollars

Source: Bank for International Settlements

Chinese Non-Financial Secotor Debt, 1995-2018 (Constant 2018 Dollars)

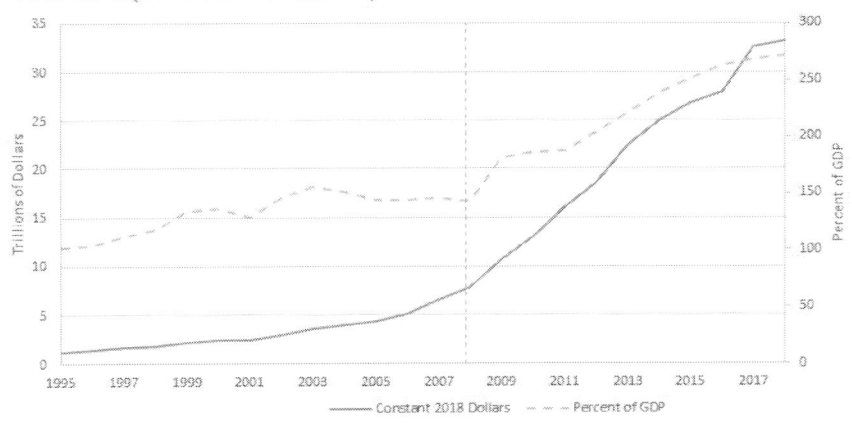

Source: Bank for International Settlements. "Total credit to the non-financial sector"; Author's Calculations
Note: Total dollar amounts were adjusted to constant 2018 dollars. Debt as a Percent of GDP was adjusted as per Chen, Wei, Xilu Chen, Chang-Tai Hseih and Zheng (Michael) Song. 2019. "A Forensic Examination of China's National Accounts" BPEA Conference Draft, Spring. Which suggested that GDP had been overstated by 1.7% since 2008.

And keep in mind that while these numbers try to take into account research showing a more accurate picture of China's – their debt has

149

likely been undervalued as well. Bloomberg reported in October 2018 that, according to S&P Global Ratings, China may have accumulated an additional 5.8 trillion in off-balance sheet debt. As reported by Bloomberg:

> *"The potential amount of debt is an iceberg with titanic credit risks," S&P credit analysts led by Gloria Lu wrote in a report Tuesday. Much of the build-up relates to local government financing vehicles, which don't necessarily have the full financial backing of local governments themselves*[100]*.*

And much of this debt may already be in trouble. Bloomberg also reported that the amount of loans already non-performing could be considerable:

> *According to some estimates, China's troubled credit could exceed $5 trillion, a mind-boggling sum that's roughly 50% of the country's annual gross domestic product. It puts China's bank stocks firmly onto a list of the most dangerous equities in the world.*[101]

Chinese Debts in Foreign Currency

Critically, China has also taken on a large amount of dollar denominated debt. Currently China's total outstanding dollar-denominated debt is estimated to be around 3 trillion USD (or roughly 27% of China's GDP).

These dollar denominated debts puts the People's Bank of China (PBoC) in a difficult position. With China's economy slowing down, China may need to further stimulate it's economy – or risk problems in its own banking system. In fact, China has devalued the yuan significantly since the Trump administration placed a 10% tariff on Chinese goods

– effectively negating the tariffs.

In April 2019, Mike Bird's article, *China's Banks Are Running Out of Dollars*, in The Wall Street Journal highlighted the fact of China's growing debt denominated in dollars, and their increasing need for dollars writing "The major Chinese commercial banks once had more dollar assets than liabilities. No longer.[102]"

In May 2019, Bloomberg published *The Dollar Dictates China's Need for a Trade Deal: The country's exploding foreign debt means it has to keep hard currency coming in.* In the article, the author Christopher Balding makes a strong case that China may be forced to negotiate a trade deal with the US because they will be unable to service their debts without bringing in dollars. He writes:

> *In 2018, China's deficit with the U.S. was equal to 92% of its entire goods trade surplus. As a matter of economics, the bilateral trade deficit has little importance – contrary to what Trump often argues. But to China, the ability to generate hard international currency is vital.*
>
> *China's external debt stood at $1.96 trillion at the end of 2018 and has probably crossed the psychologically important threshold of $2 trillion since then. The country now needs more than $100 billion annually to service foreign debtors. The pressure to cover offshore borrowing and investing in hard currency is creating an increasing shortage of dollars in Chinese banks, which have been used for a variety of political purposes*[103].

China has been able to maintain its system thus far, and may be able to maintain it longer than many expect, because unlike every other major economy, China has strict capital controls. Yuan is not able to escape China, and therefore while other countries might see capital flee the country long before the point China has reached. China has been able to prevent a collapse of its currency.

However, in May 2019 the South China Morning Post reported that Chinese banks had quietly lowered the daily limit on foreign-currency cash withdrawals - a sign that China may be feeling the squeeze of the trade war and feeling the effects of capital flight despite their controls.

Additionally, the Chinese government has been forced to run increasingly large deficits - projected by the IMF to by over 6% of GDP in 2019.

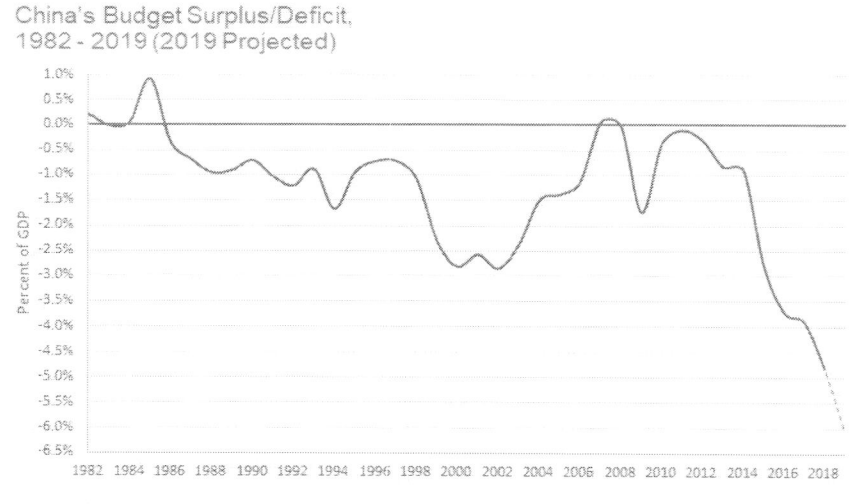

Chinese Asset Bubble

China's real estate values are extraordinarily high. Higher than Japan's during its asset bubble in the 90s[104]. China avoided the worst of the 2008 crash by adding a tremendous amount of debt and stimulus to their economy – if the bubble bursts now they could be looking at a Japanese style lost decade(s) or worse.

If China faces a financial crisis and is forced to devalue their currency substantially it would undoubtedly create massive global financial instability. Within China the cost for food imports would rise, capital would attempt to flee the country, and Chinese companies would have difficulty paying their dollar debts.

Moreover, a weaker yuan would be economically destabilizing for many other countries and geopolitically destabilizing for the world. The Trump administration is already talking about additional currency tariffs that could be imposed if China devalues their currency further.

Either way such a crisis would be unlikely to stay localized to China. Today China is the second largest economy in the world and has been almost the sole "engine for growth" in the world since the Great Recession. According to the former chief economist at the IMF, Kenneth Rogoff who says that "If there's a country in the world which is really going to affect everyone else and which is vulnerable, it's got to be China today.[105]"

China's Demographic Shift

Chinese Population by Age Group - From 1950 to 2100

Source: United Nations, Department of Economic and Social Affairs, Population Division (2017). World Population Prospects: The 2017

8

The United States

In one of the most significant geopolitical changes since the end of the Cold War, the shale oil revolution in the United States has made the US de facto energy independent. According to the New York Times, "This year [2018], the United States is expected to surpass Saudi Arabia and to rival Russia as the world's leader, with record output of over 10 million barrels a day."[106] In fact, following the shale revolution in the United States, the US now is a net oil exporter.

U.S. Crude Oil Exports (1920-2019)

Without a need to secure reliable oil supplies, and absent the existential threat from the USSR trumping all other concerns, the United States has growing skepticism about maintaining the world order that it established. Particularly with the rise of China as an economic and military competitor. The U.S. increasingly feels that the world has been somewhat free-riding at her expense. Why should the United States be providing the bulk of NATO troops and expenses to defend Europe when most nations have not contributed the agreed minimum two percent of GDP to defense for decades? Why is the United States keeping the straits of Hormuz open for oil shipments which go to China?

Having lost the status of an export economy, and no longer willing to be deferential on trade for security reasons, The US is re-evaluating the system of global trade it established through the World Trade Organization (WTO). In a 2018 report, the US trade representative argued that the US had "Erred in supporting China's entry into the WTO on terms that have proven to be ineffective in securing China's embrace of an open, market-oriented trade regime.[107]"

The US produces all the oil it needs, all the food it needs, and its economy is only slightly dependent on foreign trade – and most of this within North America . Even without a Trump Presidency, the United States would be pulling away from the rest of the world as the geopolitical reality has simply changed.

And without the United States active presence in world affairs we are returning to a world where nation states compete for resources, market access, and strategic position. Geopolitical Strategist Peter Zeihan, in his brilliant book *The Accidental Super Power,* writes:

> *For seventy years the world has not had to worry about access to markets or commodity sources. Now it will. Countries far removed from supplies of food, energy, and/or the basic matrix of inputs*

that make the industrialized world possible will face the stark choice of either throwing themselves at the mercy of superior local powers or throwing what force they can muster at the resource providers. In their desperation, many will realize that American disinterest in the world means that American security guarantees are unlikely to be honored. Competitions held in check for the better part of a century will return. Wars of opportunism will come back into fashion. History will restart. Areas that we have come to think of as calm will seethe as countries struggle for resources, capital, and markets. For countries unable to secure supplies (regardless of means), there is a more than minor possibility that they will simply fall out of the modern world altogether[108].

While the the rest of the world will suffer without American involvement, The United States is not entirely removed from the upheaval facing the world.

The United States of Debt

In the 1992 U.S. Presidential election, Ross Perot, running as an independent, took 18.9% of the popular vote. At points during the campaign he was the front-runner. His signature issue was the rising national debt. His campaign strategy was to pay for thirty minute campaign ads during prime-time television where he would discuss the economic situation. In the ads he would go into great detail - often gesturing to various graphs with his famous *voodoo stick.*

In the quarter century since, the national debt has continued to soar but it is no longer a major political issue. Certainly not an issue that could take an unknown to within reach of the presidency. In Gallup polling the issue has fallen from it's high in 1996, when twenty-eight percent of Americans said the the federal budget / budget deficit was the most important problem facing the country, to only two percent in

2018[109] [110].

Concern for the Federal budget has not dropped because of better fiscal management. Since 1996 the Federal Government has passed significant tax cuts, increased entitlement costs substantially with Medicare part D and the Affordable Health Care Act, all in addition to spending somewhere between four and six trillion dollars on wars in Iraq and Afghanistan.

Spending has largely outpaced revenue for decades, with a combination of defense spending on the Cold War and entitlement programs contributing the bulk of expenses. With the end of the Cold War, increasingly the growth of entitlement programs have driven spending.

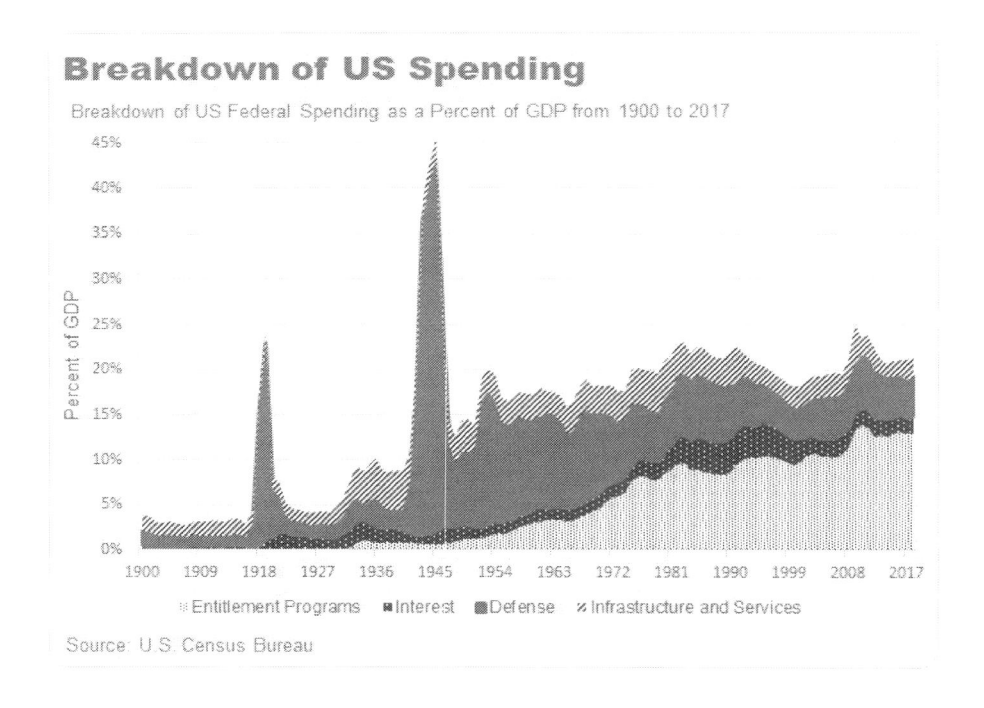

Breakdown of US Spending

Breakdown of US Federal Spending as a Percent of GDP from 1900 to 2017

Entitlement Programs · Interest · Defense · Infrastructure and Services

Source: U.S. Census Bureau

Overlaying the total revenues of the federal government as a percent of GDP shows the persistent deficits since the 1970's – as entitlement spending more than replaced defense spending.

US Spending and Revenue

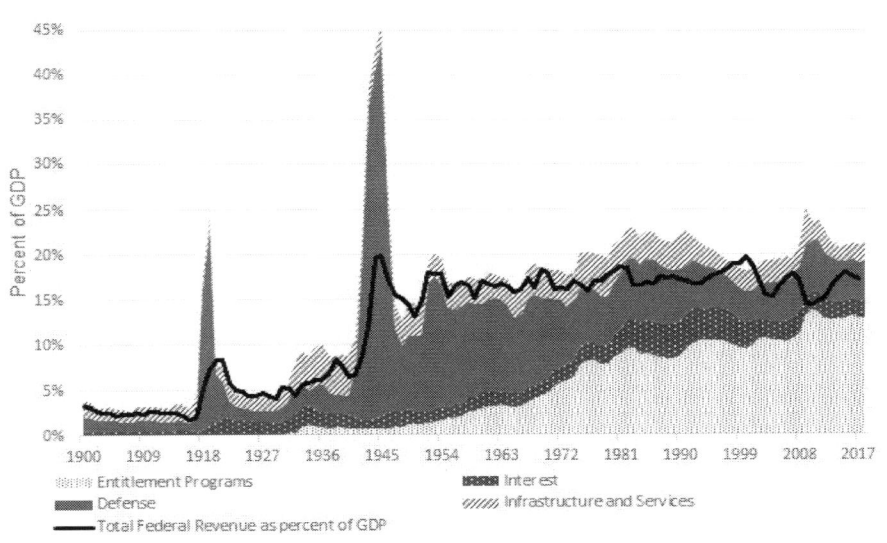

Breakdown of US Federal Spending and Revenue as a Percent of GDP from 1900 to 2017

Entitlement Programs
Defense
Interest
Infrastructure and Services
Total Federal Revenue as percent of GDP

Source: U.S. Census Bureau

When the *Baby Boomer* generation retires, government obligations for pensions and healthcare will soar. In the United States the large *Millennial* generation will eventually stabilize the tax base. But in much of the developed world there was no analogously large *Millennial* generation. In many countries the tax base will shrink dramatically concurrent with the increase in mandatory spending. Even in the United States, with relatively healthy demographics, the ratio between working-age adults and the elderly will increase significantly.

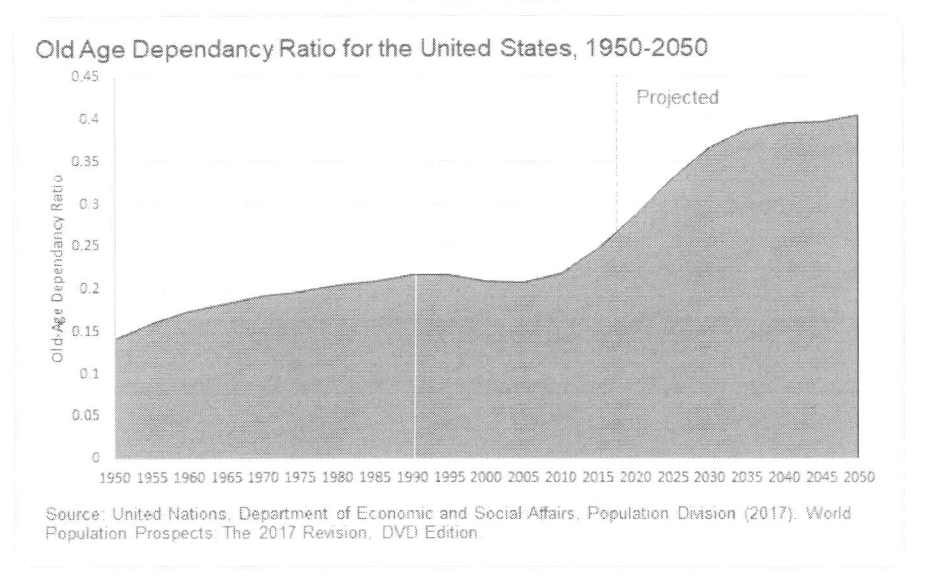

Old Age Dependancy Ratio for the United States, 1950-2050

Source: United Nations, Department of Economic and Social Affairs, Population Division (2017). World Population Prospects: The 2017 Revision, DVD Edition.

As a result mandatory entitlement spending and interest on the debt alone will take one hundred percent of tax revenues by 2019, according to CBO projections.

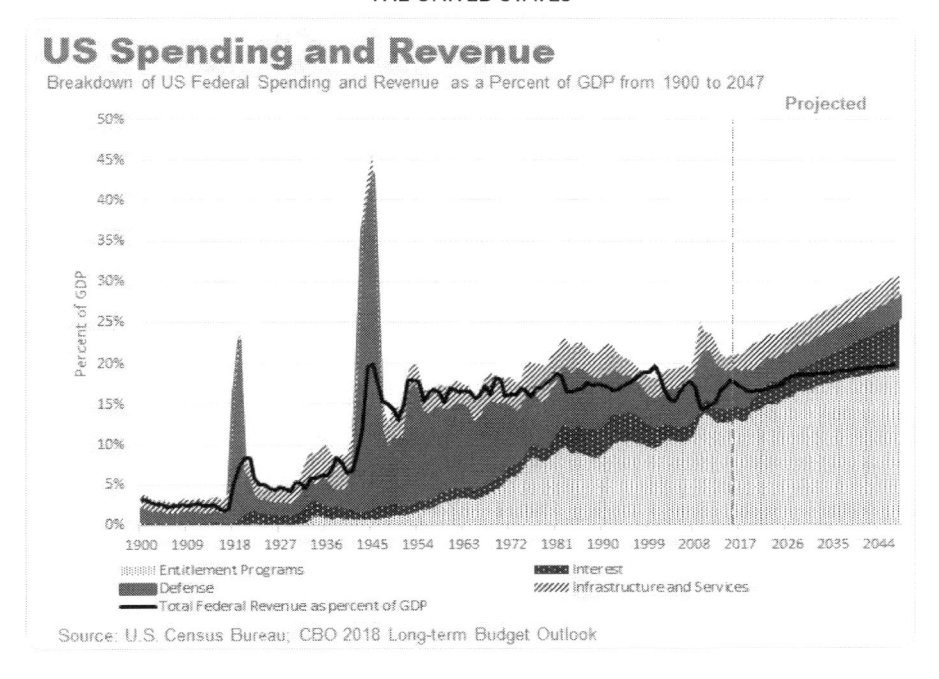

US Spending and Revenue

Breakdown of US Federal Spending and Revenue as a Percent of GDP from 1900 to 2047

[Chart showing US Federal Spending and Revenue as percent of GDP from 1900 to 2047, with categories: Entitlement Programs, Defense, Total Federal Revenue as percent of GDP, Interest, Infrastructure and Services]

Source: U.S. Census Bureau; CBO 2018 Long-term Budget Outlook

Scale of the United State's Debt

The CBO estimates that to reduce the United States Federal debt to it's fifty year average of forty-two percent by 2048 the government would need to cut spending or raise revenues by 3.0 percent of GDP each year – for the next ~ **thirty years.** To put that in perspective, it's just slightly less than cutting the entire U.S. defense budget. Equivalent to a seventeen percent increase in revenues or a fifteen percent cut in spending – again maintained for thirty years.

Crucially, these estimates are if the changes took place by 2019. To have the same effect a decade later in 2029, 4.5 percent of GDP would need to be cut from spending or raised from revenue. Almost double the projected amount spent on all defense spending- or a twenty-five percent increase in revenues. It seems extremely unlikely that the current Trump administration, or any possible successor in 2020, would be advocating for a generation of higher taxes and lower spending.

Optimistic Nature of the Projections

All of these estimates are base on the official CBO estimates and projections. However, these CBO projections forecast no recessions or substantial increases in interest rates in the foreseeable future. Also note how estimates for future defense spending is almost cut in half in the projected future. This is not because the defense budget is going to be slashed but an accounting trick perpetrated by the congress. Rather planning on increasing the military budget, they plan to cut it and then simply issue a wavier every year. In fact, military budgets around the world, including in the United States, are booming[111].

Another accounting trick, mandated by congress, is ignoring fair-value accounting. Fair value accounting takes into account the risks to repayment. For example, currently the CBO is mandated to treat all loans as having the same risk as Treasury Bonds - essentially no risk. A mandate that forces the CBO to estimate, for example, that the Federal Student loan program will bring in a 184 billion dollar profit over the next decade. While fair-value accounting by the CBO would project a 95 billion dollar *loss.* The CBO has long been in favor of the more realistic fair-value approach, recently arguing that:

> *In CBO's view, a fair-value approach would provide a more comprehensive measure of the costs of federal credit programs to the government than what is currently reported because it would fully incorporate the cost of* **market risk**.[112]

In a recent edition of *National Affairs,* Jason Delisle and Jason Richwine weigh in on the accounting debate. Arguing, in *The Case for Fair-Value Accounting*, that:

> *In nearly every government credit or insurance program—from student loans to public pensions to green-energy subsidies—the government makes risky investments without recording the full price of the risk. The government, at both the federal and state levels, exploits this accounting flaw to generate dishonest financial schemes—and some would even be illegal if they occurred in the private sector*[113].

Even with all of these accounting gimmicks, rosy economic outlooks, and the predicted perpetual low-interest rates, the budget deficit is projected to be an incredible ten percent of GDP by 2047, only a few percentage points away from the level that sparked Greece's crisis. Moreover, total debt will rise to 180% of GDP in the CBO's baseline projection -with little hope that things will improve beyond the end of the projection period.

Debt Rollover

Despite the bleak fiscal picture, it is difficult to even imagine that the United States could face a true fiscal crisis. The very concept seems unreal. However, as Reinhart and Rofoff have detailed in their seminal work *This Time is Different*, where they studied eight centuries of sovereign debt defaults and inflation crises, massive built ups of sovereign debts followed by defaults are not rare in a larger view of history. As they note:

> *All too often, periods of heavy borrowing can take place in a bubble and last for a surprisingly long time. But highly leveraged economies **particularly those in which continual rollover of short-term debt is sustained only by confidence in relatively***

***illiquid underlying assets**, seldom survive forever particularly if leverage continues to grow unchecked. This time may seem different, but all too often a deeper look shows it is not[114]; [emphasis added]*

One important issue that adds to the risk of a debt crisis in the U.S. is the short-term nature of the U.S. debt burden. In order to get even lower interest rates on the U.S. debt, the Treasury department relies heavily on short term rather than long term debt. Continually rolling over massive amounts of debt every year. Steven L. Schwarcz writing in, Rollover Risk: Ideating a U.S. Debt Default, explains:

The U.S. government relies heavily on short-term debt funding. As a result, some estimates suggest that the U.S. government has to roll over half of its debt every two years. A recent article estimates that the U.S. government will have to roll over seventy-one percent of its privately held debt over the next five years. Additionally, the government depends on short-term debt to finance the federal budget deficit—which was $1.1 trillion in 2012. So long as the United States maintains a large deficit funded by the issuance of short-term debt, it creates a risk of being unable to raise the funding necessary to avoid default.

While the government saves hundreds of billions of dollars by rolling over short-term debt rather than using longer year bonds - this creates the risk of a political stalemate over the debt ceiling or a interest rate shock making it impossilbe for the U.S. to rollover the national debt. It was this risk that led S&P to downgrade the U.S. AAA credit rating to AA+ in 2011.

Despite all of this debt, the United States is uniquely positioned among the nations of the world to benefit from a restructuring of the global economy. The United States has easily the most healthy demographic picture among the great powers. Especially when compared to rivals like China and Russia. China and Russia are also significantly more dependent on globalization the continuing economic trends of the past half-century. The United States can only benefit from the end of trade deficits.

With the gains in shale oil production having made the US energy independent, the US no longer needs to maintain stability in distant nations, or be subject to potential shocks in the world oil markets. In strong contrast to Russia and China who are dependent on exporting and importing oil respectively.

The United States is not only more insulated economically - but it is insulated geographically. Any political or geopolitical instability will not effect the United States in the same way that Europe and Asia are impacted. Not only could heightened geopolitical tensions cause oil shocks which they will need to deal with - but the United States has two oceans which insulated it from the type of destabilizing migrations that took place in Europe following the war in Syria.

Article in Foreign Affairs, *With Great Demographics Comes Great Power*, by Nicholas Eberstadt

> *Compared with other developed countries, the United States has long enjoyed distinctly high immigration levels and birthrates. Between 1950 and 2015, close to 50 million people immigrated to the United States, accounting for nearly half of the developed world's net immigration over that time period. These immigrants and their descendants made up most of the United States' population growth over those decades. But U.S. fertility is also unusually high for an affluent society. Apart from a temporary dip during and immediately after the Vietnam War, the United States' birthrates after World War II have consistently exceeded*

the developed-country average. Between the mid-1980s and the financial crisis of 2008, the United States was the only rich country with replacement-level fertility. Assuming continued levels of immigration and near-replacement fertility, most demographers project that by 2040, the United States will have a population of around 380 million. It will have a younger population than almost any other rich democracy, and its working-age population will still be expanding. And unlike the rest of the developed world in 2040, it will still have more births than deaths.

All of this is not to say that the United States does not face major risks during this period of upheaval. Only that the odds are stacked in the United States favor, and unlike many other nations, the United States also has much to gain.

IV

Financial Crises and The Sovereign Debt Crisis

9

The Sovereign Debt Crisis

Karl Marx believed that instability was an inherent feature of capitalism; that the economic booms and busts of the business cycle would not only continue, but they would accelerate - worsening with each cycle until the capitalism system itself imploded. Over the last one hundred and seventy years, capitalism has not failed. In sharp contrast to a host of failures and misery wrought by attempts at communist societies, free-markets have led to a world of previously unimaginable wealth and prosperity. However, market economies *have* been plagued with a continual cycle of booms and busts and frequently this economic turmoil leads directly to political turmoil.

Supply and demand should keep the economy in balance, yet with frustrating regularity we find that some assets have been overvalued. Often stocks or housing is the culprit - and everything from *animal spirits* to excessive savings has been put forth as a cause. While financial crises like the Great Depression have played an enormous role in our history, they can seem almost inexplicable. Farmers will let crops rot in the fields while people starve in cities. Factories making essential goods are shut down, and the workers laid off. Large percentages of the workforce remain idle while people are desperate for work and for the goods operating factories would provide. The smooth functioning of the market seems to breakdown.

To understand these crisis it is important to understand that there is a systemic cause. Although it is certainly possible to have an economic crisis because of an unexpected shock - war, invasion, plague etc. – this is rare. Most are not caused by an external shock, but follow a predictable pattern of boom and bust. This book began with a synopsis of the 2008 Financial Crisis, And thinking about the crisis you might say that speculation in housing or lax regulation in the financial system was cause of the crisis. And while those points are true - it misses the larger reason for the crisis. As Hyman Minsky explained talking about financial crises:

> *Once the sharp financial reaction occurs, institutional deficiencies will be evident. Thus, after a crisis, it will always be possible to construct a plausible argument – by emphasizing the trigger events or institutional flaws – that accident, mistake, or easily correctable shortcomings were responsible for the disaster*[115].

But, as Minsky went on to explain in further detail, these are themselves symptoms are a more systemic problem. To understand what causes this pattern of boom and bust - the flaw in the heart of the free market - it is necessary to understand the workings of the banking system. To understand money and prices and how their manipulation can cause havoc in an economy.

* * *

The Money Supply, Prices, and Resource Allocation

The concept of money serves an enormously valuable function for our society. We only have to look at times and places where it stopped functioning, such as Germany in 1923 or Venezuela today, to see how civilization itself breaks down in its absence. However, currencies can malfunction in ways other than hyperinflation. A disruption in the pricing mechanism of currency also causes recessions and depressions.

Money allows goods to be priced, and allows for the calculation of profit and loss. Profit and loss enables the economy to make the most efficient use of scarce resources. Should more steel or aluminum be manufactured? More SUVs or bicycles? Does it make sense for a company to hire another employee? Such questions are impossible to calculate in the aggregate[116], but by looking the profitably of such choices, individuals can make decisions using the distributed knowledge of everyone in the economy. This is the core idea of the market economy and the *invisible hand* that directs resources. Resources flow to where they are most effective and therefore most profitable. Profitability calculations allows production levels to match the needs of the ultimate consumers. Profitability allocates resources efficiently because it is based on market prices, which are ultimately based on supply and demand.

Changes to the Money Supply

It is a basic tenant of economics that interference with market prices, such as price controls, will distort profit-and-loss calculations and lead to misallocation - shortages or gluts of goods. Price controls will cause resources throughout the economy to be mis-allocated. Although not as intuitive, changes in the money supply will also distort prices and lead to a similar misallocation of resources. When discussing the money supply and prices, it is important to differentiate between *price* inflation, and inflation *of the money supply.* While an increase in the money supply will distort prices and lead to resource misallocation, it may not always cause a general increase in prices. Moreover, when prices rise, they are unlikely to all rise evenly. Often one type of good or asset will rise substantially higher than others. This is because of the time it takes for the market to adjust to the higher level of total currency, and because typically an increase in the money supply is not distributed evenly.

Increase in the Money Supply

A simple thought experiment can illustrate the impact of increases and decreases of the money supply and their relationship with prices. Imagine the scenario where everyone in the world has the number of dollars they own doubled overnight. The world is not any richer. The total number of goods and services in the economy is same. What would happen is that prices would roughly double. A soda that cost one dollar before would cost two dollars. However, because of the time it would take for the market to adjust prices to this new level of currency, whoever spent their money first would get the most *bang for their buck*. The value of savings would be cut in half when prices adjusted. One dollar in savings would only buy half of what it could before. The total value of the goods and services in the world would be unchanged - but value would have been transferred from savers to those who spent the new money before prices adjusted.

This change in the money supply would not only transfer wealth, but it would also cause a misallocation of resources until all the prices in the economy reached a new equilibrium with the larger supply of currency. With double the amount of money in their pockets, and prices not yet adjusted, people would spend more on consumption - triggering a boom. Business' profit would rise as their sales would go up before their costs did. This rise in profit would cause businesses to misallocate resources. More resources would shift into producing consumer goods than are warranted. The increase in profits was not truly a reflection of a change in desire for more consumer products. The change was a reaction to the distortion of the value of money itself – distorting the allocational mechanism of profit and loss. Business would see profits fall as consumer demand returned to normal levels as production costs adjusted. In fact, following the boom there would be a bust as the previous artificial increase in profits would lead to an oversupply of some goods. Workers would be laid off as the economy shifted resources back to their original uses based on true consumer demand.

Decrease in the Money Supply

Now imagine the opposite scenario. The number of dollars that everyone has is cut in half. In time prices would adjust to the new levels. A soda that cost one dollar would sell for fifty cents. Here though savers would benefit as the value of their savings would double in real terms. Businesses would suffer as sodas they had purchased for seventy-five cents to sell for a dollar are now only worth fifty cents. Rather than a surge in sales and demand outstripping supply, you would see a glut of unsold goods –and a sudden fall in business' profits. In time their costs would also fall and they would return to their former level of profitability. Assuming they could take the loss on their inventory and stay in business.

While these thought experiments are helpful in illustrating the impact of changes in the money supply, they are not very representative of the economy. The currency does not generally double overnight. Nor are changes in the money supply evenly distributed among consumers. For a more realistic picture it is important to realize that the interest rate is another price, determined between the supply of savings and the demand for borrowing which allocates resources in the economy between future consumption and current consumption. The more money that is saved, the larger pool of resources available for individuals or business to borrow. Interest is the price that is paid to someone for them to forgo their current consumption so that the borrower can use it immediately.

Fractional Reserve Banking

It is also important to understand how changes to the money supply take place through the banking system. The world's banks operate under *fractional reserve banking.* In this system, banks hold only a fraction of their reserves in liquid assets, and yet lend out a multiple of that amount. This practice of lending out more money than they have in deposits increases the total amount of money in the economy. What this means is that for every dollar in assets a bank has, it can increase the money supply by lending out a multiple of that amount. Likewise, a bank writing off a bad loan lowers the total amount of money in the economy. If a bank has a capital requirement of three percent, then for one hundred dollars in assets a bank can lend out thirty-three hundred dollars. The reverse is also true because losing one hundred dollars in reserves means it must decrease its leverage by thirty-three hundred dollars.

Through this mechanism of new loans adding money to the economy, the total money supply can be increased. Central banks attempt to control the money supply by adjusting the interest rates they charge to banks. If a central bank wants to increase the money supply, it will lower interest rates. This will increase the amount of bank loans taken out by people and businesses. Such lending adds currency to the economy - increasing the money supply. If the central bank wanted to decrease the money supply, or more likely to decrease the *rate* of the increase, (perhaps because the rate of price inflation was becoming too high, or because it suspected a bubble of mis-allocated resources had developed) it would raise interest rates. This would encourage banks to leave their money with the central bank and make it harder to borrow. This would tend to decrease the amount of loans taken out and so also the rate of inflation.

By manipulating interest rates, the amount of loans taken out through the banking system, and thus the money supply, the economy can be controlled by the central bank. However, unlike in the earlier thought experiment, bank loans are not distributed among the population evenly, and the increase in the money supply does not affect all goods equally. By lowering interest rates, goods that are most sensitive to interest rates are the most likely to be over-invested in.

Economic Bubbles and Fractional Reserve Banking

The 2008 housing bubble is a classic example of artificially low interest rates causing over-investment and raising prices for a specific interest-rate-sensitive asset. But the booming economy of the Roaring Twenties was likewise the result of artificially low interest rates. Barry Eichengreen, author of several books on monetary history including *Golden Fetters*, an excellent look at the monetary policies of the Great Depression, writes that:

> *Accounts of the twenties in the United States (such as Kindleberger 1973) emphasize the ready availability of credit, reflecting the ample gold reserves accumulated by the country during World War I, the stance of Federal Reserve policies, and financial innovations ranging from the development of the modern investment trust to consumer credit tied to purchases of durable goods like automobiles. Credit fueled a real estate boom in 1925, a Wall Street boom in 1928-9, and a consumer durables spending spree spanning the second half of the 1920s.*[117].

The misallocation of resources from the artificially low interest rate created the housing bubble and the stock market bubble of the 1920s. Which was brought to an abrupt end when the Federal Reserve, alarmed by the speculation in the stock market, raised rates (lowering the money supply) in 1928.

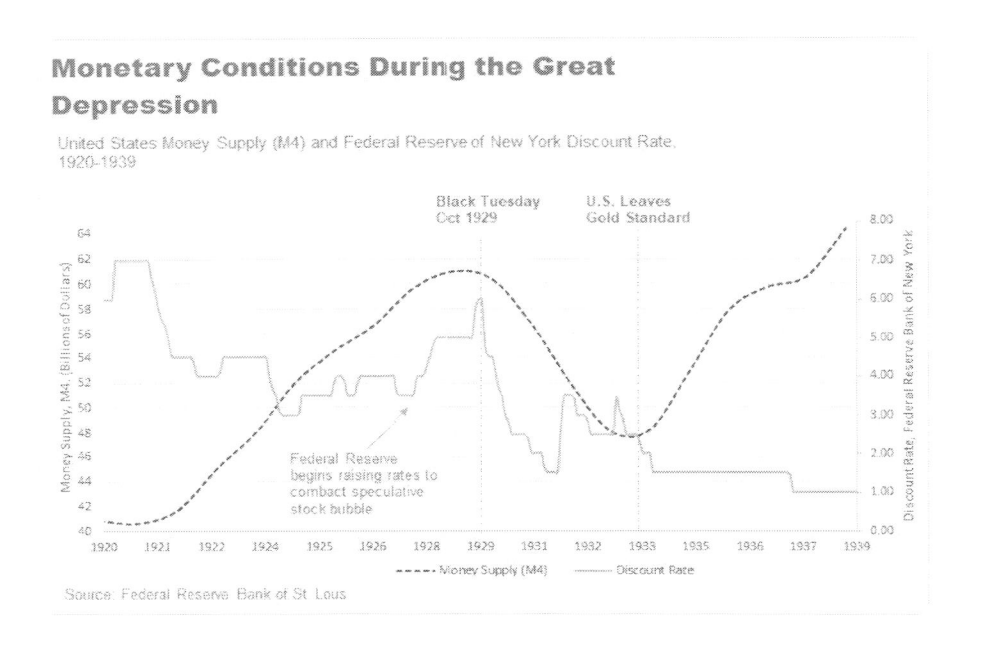

Monetary Conditions During the Great Depression

United States Money Supply (M4) and Federal Reserve of New York Discount Rate, 1920-1939

Source: Federal Reserve Bank of St Louis

Deflationary Spirals

However, this misallocation alone is not what made the Great Depression as deep and prolonged as it was. Fractional reserve banking allows the system to expand the money supply - transferring wealth and creating bubbles of misallocated resources. But it also creates the possibility of a deflationary spiral. When banks are heavily exposed to an asset, such as housing or stocks, which have become overvalued and the bubble bursts- the banks losses could be so great they lose all of their reserves. In that case depositors, which were never invested in those assets, can lose their money.

Imagine a bank that has one hundred dollars in deposits, and suppose it has lent three hundred dollars to borrowers keeping only ten dollars in reserve. But the bulk of these these reserves are not cash – they are in safe assets like mortgages. If these fall in value sufficiently the bank will not have enough funds to maintain operations and will be forced to close. Or if ten percent of depositors lose confidence in the bank and attempt to withdraw ten dollars in deposits from the from the bank will run out of money. In such cases the bank is forced to sell all of its assets, and the one hundred dollars in deposits simply evaporates. This lowers the total money supply in the economy, and puts more pressure on the remaining banks in an ever increasing deflationary spiral.

Deflation During the Great Depression

During the Great Depression bank deposits were not insured by the federal government at all. Several banks had loaned heavily to investors in the stock market and were devastated by the losses. One such case, similar to Lehman Brothers in the 2008 crisis, was the Caldwell Company. Economist Gary Richardson of the Federal Reserve Bank of Richmond explains that:

> *Caldwell was a rapidly expanding conglomerate and the largest financial holding company in the South...The parent got into trouble when its leaders invested too heavily in securities markets and lost substantial sums when stock prices declined. In order to cover their own losses, the leaders drained cash from the corporations that they controlled.*
>
> *On November 7, one of Caldwell's principal subsidiaries, the Bank*

of Tennessee (Nashville) closed its doors. On November 12 and 17, Caldwell affiliates in Knoxville, Tennessee, and Louisville, Kentucky, also failed. The failures of these institutions triggered a correspondent cascade that forced scores of commercial banks to suspend operations. In communities where these banks closed, depositors panicked and withdrew funds en masse from other banks. Panic spread from town to town. Within a few weeks, hundreds of banks suspended operations[118].

Concerned by these bank failures, people throughout the country removed their deposits from banks. However, the nature of the fractional reserve system is that banks have loaned out depositor's currency – and then some. With mass withdraws from the system, banks were forced to shut down. The effect of this was to dramatically reduce the money supply in the economy. Transmitting the crisis from Wall Street to Main Street. Each bank that shut down, and each person who withdraw their funds from the banking system reduced the leverage of the banks and thus the total money supply. As in the earlier thought experiment, this dramatic decrease in the money supply created even more issues for the economy. With less supply of dollars the value of the dollar increased, goods became overpriced and businesses needed to sell them at a loss. Business throughout the economy slowed down – or came to a halt. Unemployment soared. The spiral of deflation did not end until the nation left the gold standard and inflated the currency.

Deflation During the 2008 Financial Crisis

The existential risk to the banking system is a deflationary spiral. This is what almost happened in the 2008 financial crisis. Although it was halted before it reached such extremes as seen in the Great Depression. In 2008 Subprime mortgages were being held by banks as assets as part of their capital reserves. The money supply expanded as banks created and lent out dollars based on multiples of that capital. This addition of new money into the economy - an increase in the total amount in circulation - created a boom. But it also mis-allocated resources, creating over-investment in interest-rate-sensitive assets, particularly in housing. In time it became clear that subprime mortgages were not worth the amount that banks had valued them at. As housing had been over invested in, prices fell to reflect that. Subprime mortgage borrowers defaulted on their loans.

This forced banks to sell assets to meet their capital reserve requirements. This simultaneous sell-off lowered the value of their remaining assets in a vicious cycle. As banks were desperate to raise capital, lending was frozen, and credit contracted. Much like the failure of the Cadwell Company in the Great Depression, The collapse of Lehman Brothers created uncertainty throughout the system. Monetary and price deflation set in worldwide as the monetary base shrank from the de-leveraging of the banking system.

US Money Supply Level

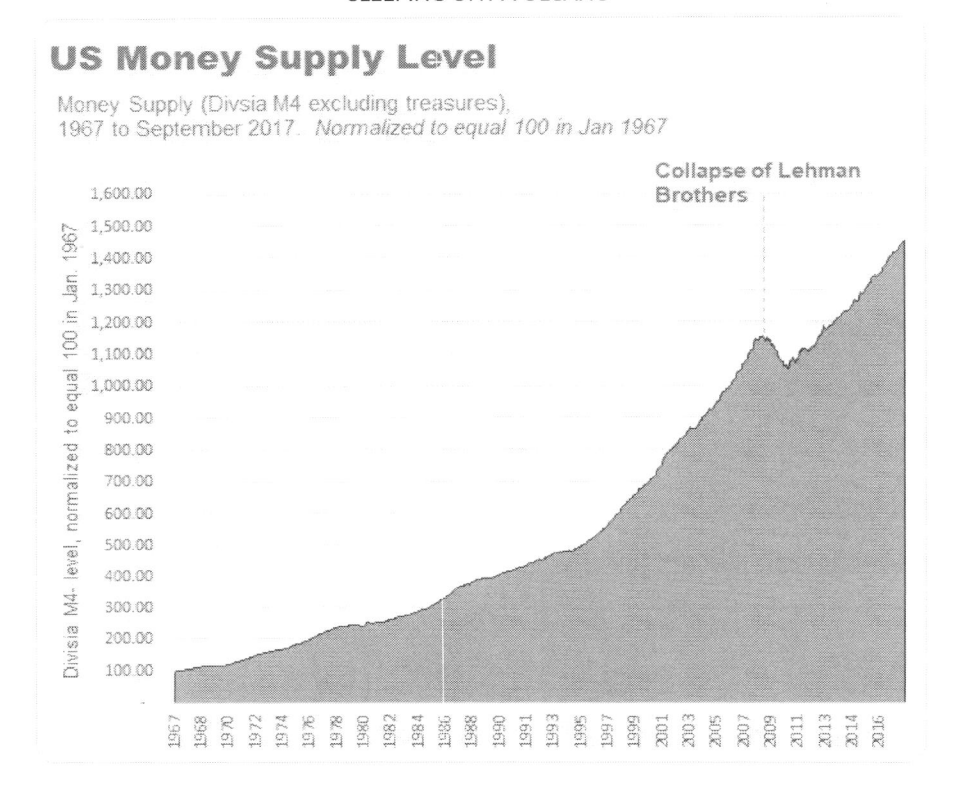

Money Supply (Divsia M4 excluding treasures),
1967 to September 2017. *Normalized to equal 100 in Jan 1967*

[Chart: Divsia M4 level, normalized to equal 100 in Jan. 1967, from 1967 to 2016, showing steady growth with a dip labeled "Collapse of Lehman Brothers" around 2008, rising to approximately 1,500 by 2016.]

This deflation affected not only the profit of banks – but the profit of all companies in the economy. The reduction of the money supply increased the value of dollars. This lead to a fall in prices and thus a fall in profits for all inventory – even in sectors which may not have experienced over-investment. The world prevented economic Armageddon by having central banks re-capitalize the struggling financial institutions. Bailing out the struggling institutions by buying the "toxic assets" at their previously high values – short-circuiting the deflationary cycle – although this came at the cost of increasing the sovereign debt of governments around the globe.

It is important to realize that it is the economic boom from increasing the money supply which is harmful to long-term growth. The correction, the reversion to actual supply and demand, is beneficial in the long-run. With the extremely large caveat that there is the real risk of a deflationary spiral that goes too far and risks the existence of the banking system itself. It is this inherent instability in the fractional reserve banking system that complicates the retirement of the Boomer generation and the unsustainiblity of sovereign debt.

Volcanic Eruption

The sudden and unprecedented fall in the working-age population of the world, along with the aging of the Baby Boomer generation, will be an immense shock to the global economy. The massive size of the Baby Boomer generation and the *demographic dividend* of lower fertility rates has driven much of economy over the last fifty years.

The inevitable shock of the reversal of this age structure has been worsened through decades of governments taking advantage of the size of the Boomer generation to maintain a raft of unsustainable polices. Policies whose flaws have been papered over by the massive and disproportionate size of the Boomer generation. Tremendous wealth which could have been used productively was instead not only spent frivolously - but used as collateral to build massive debt burdens. As the former chief economist of the BIS lamented - "I am afraid that at some point this is going to be resolved with a lot of debt defaults. And what did we do with the demographic dividend? We wasted it." Indeed, borrowing from the Boomer generation has become essential to balancing budgets for nations and corporations around the globe. The world has become addicted to debt. Debt has risen to levels only possible by the disproportionately large Baby Boomer Generation and the artificially low interest rates maintained by central banks.

The global economy is primed for a shock not only because of the shift in the age structure of the World. The failure of the Bretton Woods system in 1971 exacerbated the distortion considerably. It enabled a massive distortion in exchange rates, and, as Paul Volker, the renowned Fed Chair who tamed inflation in the late 1970s, said:

A nation's exchange rate is the single most important price in its economy. It will influence the entire range of individual prices, imports and exports, and even the level of economic activity[119].

When this exchange rate distortion was combined with the post-war geopolitical environment and the Baby Boom, the effect was to shift a substantial amount of trade and economic activity in an unsustainable way. Creating massive trade deficits while substantially increasing the national debt of many countries. Central banks have attempted to mitigate the increased economic instability and the servicing costs of the massive debt burdens by lowering interest rates lower than ever before. But this has only delayed restructuring and caused additional misallocation of resources. These artificially low rates and manipulated currency exchanges have created mass distortion in the global economy. The link between supply and demand and profit and loss has been mangled. The massive restructuring needed will be painful, and the inability to pay the debt that has already been accumulated more painful still.

However, the real risk to the world is not merely that governments will be unable to pay their debts - or that taxes will have to rise and entitlement programs cut.

The Risk from Debt

Sovereign debt underpins the global financial system. It is considered the safest of all assets - legally assigned zero risk of default in some nation's banking systems. If sovereign debt fell in value like subprime mortgages assets did in 2008, the world would face a financial crisis that would make The Great Recession pale in comparison.

In 2008 mortgages lost value and banks were forced to deleverage – creating a deflation of the money supply that caused havoc throughout the economy. But after missteps with Lehman Brothers, the government stepped in to prevent a spiral of deflation. During the Great Depression, stocks lost value – similarly creating deflation until the government ended the gold standard and stepped into increase the money supply and stop the spiral.

In addition to the risk of missteps or costly delays in reacting to halt to a deflationary spiral arising from a sovereign debt crisis, sovereign debt has additional inherent challenges. With sovereign debt, it is more difficult for the government to intervene. As Mervyn King the head of the U.K. central bank said during the 2010 sovereign debt crisis in Southern Europe:

> *Dealing with a banking crisis was difficult enough, but at least there were public-sector balance sheets on to which the problems could be moved. Once you move into sovereign debt, there is no answer; there's no backstop.*

Moreover, consider the integration of the global banking system. Many relatively healthy nations are reliant on the debts of other weaker ones. Spanish banks have lent the equivalent of six percent of Spain's GDP to Turkish banks . Spain in turn owes hundreds of billions to French and German institutions . Moreover, nations cannot support the value of government debt the way they did with toxic mortgage debt. Nations of the world risk a *Sovereign Debt Doom Loop* -a spiral where the solvency of a nation itself is called into question. Any purchasing to prop up the value of the debt only serves to further weaken the fundamentals of the country. This issue was highlighted in an article in Bloomberg by the

Editorial Board, criticizing the failure of the new post-Financial Crisis Basel III bank capital regulations to address the issue of sovereign debt. As the article notes:

> *National regulators can choose to treat government bonds denominated in domestic currency as safe — and they do. All the committee members use this freedom to set the risk weight on such bonds at zero. This means their banks can load up on domestic sovereign bonds without needing to raise any more capital.*
>
> *As a result, some banking systems serve as major creditors to their respective governments. According to a study by the Bank for International Settlements, government debt represents more than 10 percent of banks' assets in countries such as Spain and Japan. In Italy, it's nearly 20 percent.*
>
> *These holdings pose a danger. If investors turn against those bonds, they might inflict severe losses on the banks concerned. If governments then have to step in, issuing more debt to meet the cost, the value of the bond holdings might fall again as the governments' creditworthiness is called into question — the so-called "doom loop" that can turn a banking upset into a fiscal and financial crisis*[120].

The economic impacts of the demographic shift make it likely that many governments, and all but inevitable that some, will be unable to service their debt. The interdependent and leveraged nature of the banking system, and it's reliance on sovereign debt which is considered virtually risk free by regulators, means that even strong economies are at risk of contagion and face an existential threat to their economic systems.

This is not to say that nations have no options to deal with a sovereign debt crisis on a large scale. However, they are not pleasant options. If the world loses confidence in sovereign debt, the nations of the world face will face an impossible choice. Harsh and politically infeasible austerity to pay down their debts (which is likely to backfire as it has in Greece); defaulting or restructuring the debt, which would be incredibly deflationary (leading to prolonged and deep recessions); or monetization of their debts leading to high inflation and intensifying competitive currency devaluation between nations.

Default and Deflation

Although the worst case scenario deflationary spiral not the most likely scenario, it is worth understanding what could happen.

For example in the United States, if confidence was lost in US treasuries, interest rates would rise. Investors would need more interest to compensate them for the risk. This would only make it harder for the nation to pay the interest on the debt. Moreover, with rising interest rates, the housing market, corporate debt, and zombie companies would be in serious trouble. Again, worsening the nation's fiscal position by reducing tax revenues. Higher interest rates would also put the banking system in a bind as defaults would spike. In such a scenario – similar to 2008 - the US government might see its hands tied to calm the markets. If the government stepped into prop up the housing market or the stock market – taking on the bad debt, investors would become *more* concerned with the fiscal health of the United States. If the government intervened rates would rise higher - necessitating another yet government response. This is a fiscal crisis or a *Sovereign Debt Doom Loop*. As the panic builds, the government is unable to stop the carnage because it is the debt of the nation itself that is the root of the crisis.

To imagine what that might look like it is worth remembering the warnings that Ben Bernanke and Henry Paulson gave to the congress in 2008 if action was not taken to stop the spiral of deflation. The banking system would cease to exist; depositors would be unable to withdraw money from their accounts. Many large businesses would immediately be unable to make payrolls. There would be the real risk shipping would breakdown and halt the flow of food to stores. The government, as Paulson warned in 2008, would need to plan for martial law and how to feed the public. The economy could simply collapse.

Without ending the spiral, the nation could find itself in a decade or longer depression that would no doubt lead to political turmoil and unpredictable chaos.

Austerity

Another option is austerity in an attempt to convince investors that the debt can be repaid and that they should maintain confidence in the system. Greece is the poster child for this type of austerity. Deep cuts to pensions, healthcare, substantially higher taxes – all with more of the money simply going to pay the now higher interest rates demanded to service the debt. Over the last decade Greece has been mired in a depression deeper and longer than the Great Depression while their debt to GDP ratio has only increased.

In the United States such a course of action might hypothetically work. The millennial generation was large enough that deep cuts to entitlements and substantial tax increases could put the United States in a position where it could pay off its debt. But this approach seems highly unlikely if only because of how politically unpalatable such an approach would be. Who would vote for a generation of sustained punishing taxes and unconscionable cuts to healthcare and pensions? Likely while in the midst of a deep recession. We have seen the wave of populist anger following the economic hardship of the Great Recession and it is difficult to imagine that such a plan, even if enacted, could be maintained.

Keep in mind that tax increases on the rich alone would be totally insufficient to balance the budget. Extrapolating from the CBO's estimate on the impact of raising taxes on the top two income brackets – even if tax rates were increased to 100% (and assuming that 100% tax rates would not increase tax avoidance-legal or otherwise) – the deficit would only be cut roughly in half. If, rather than cutting spending, higher taxes alone were the solution it would require a dramatic and sustained increase in tax revenue. Which would have to fall fairly substantially on the middle class. And this, again, is only a possibility for the United States. For Japan, Southern Europe, Germany, and many other countries demographics make this less of a viable option.

There is no political will to cut spending and pay down debt, and for many nations the demographic outlook makes it pointless to try. Moreover, once an economy has fallen into recession, austerity measures can be counterproductive. If a nation falls into recession and austerity is continued – it risks falling into the very deflationary spiral that it sought to avoid.

Monetization and Inflation

Given the choice between likely counterproductive austerity, default, or monetization, monetization of the debt is perhaps the most likely outcome. If only because it is the path of least resistance politically. No politician in a democracy would be likely to get elected pushing for brutal tax increases and cuts to social services. Every economist realizes that default on such a scale would risk the collapse the entire economic system and would plunge nations into economic depression. Monetizing the debt is more politically palatable and more justifiable economically. Every pensioner still gets paid, no spending is cut, no taxes are raised – inflation just rises a bit. In nations like Japan this could even be spun positively initially.

The issue is that the government would be unable to stop the inflation. The huge budget deficits are projected to continued indefinitely and therefore must be paid for with new printed money indefinitely. Eventually inflation rates would pick up and nations options have now been even further limited. If nation's stop adding to the money supply to make up the deficit, then they will be forced to either default on the debt – which would be deflationary and risks collapsing the economy – or simply instantly cut a huge swath of the federal budget with politically intolerable austerity. And of course, such an immediate and substantial drop in government spending would itself be deflationary and a tremendous shock to the economy.

Moreover, once inflation rates and inflation expectations pick up, the government itself will have to pay more interest on it own massive debts as interest rates would rise to compensate for the inflation. Writing in *The Future of Public Debt: Prospects and Implications*, BIS economists Stephen G Cecchetti, M S Mohanty and Fabrizio Zampolli detail the potential inflationary impact, and how massive debts tie the hands of central banks:

> *When the public reaches its limit and is no longer willing to hold public debt, the government would have to resort to monetisation. The result, consistent with the quantity theory of money, is inflation. And anticipation that this will happen may also lead to an increase in inflation today as investors reassess the risk from holding money and government bonds.* **In such an environment, fighting rising inflation by tightening monetary policy would not work, as an increase in interest rates would lead to higher interest payments on public debt, leading to higher debt, bringing the likely time of monetisation even closer. Thus, in the**

absence of fiscal tightening, monetary policy may ultimately become impotent to control inflation, *regardless of the inflation fighting credentials of the central bank*[121]. *[emphasis added]*

In 1980, Paul Volker at the Federal Reserve took the Federal Funds target rate to almost twenty percent in order to curb inflation. While such an interest rate would be unimaginable today and devastating to the corporations and households who rely on low interest rates, it would also ultimately be ineffective against inflation. Today, the government itself is so reliant on debt that a spike in interest rates would increase the amount of money the government would need to print to service the debt.

William White, the former chief economist at the BIS is also concerned about the ultimate impact of sovereign debt on inflation. In 2018 he warned that:

There is a significant risk that this is going to end badly because the Bank of Japan is funding 40pc of all government spending. This could end in high inflation, perhaps even hyperinflation[122]

When the head of the BIS, a prestigious and level-heaved central banker, who correctly predicted the 2008 housing crisis begins warning the world of possible hyperinflation in major economies the warning should be taken seriously.

What is Now Inevitable

It is impossible to know exactly how the issues with sovereign debt will unfold. Nations have been able to use quantitative easing (QE) to keep the rates on their debt artificially low, and thus make payments possible . Perhaps Japan will run out of eligible debt to purchase for QE and rates on Japanese government bonds (JGBs) will rise leaving them unable to make interest payments. As early as 2010 there were concerns about both the EU and Japan running out of eligible bonds for QE. Maybe internal political issues in the eurozone will prevent the ECB from supporting a nation like Italy or Spain leading them to leave the euro and default - sparking a wave a contagion and defaults. There is the risk that China could monetize it's debts by dramatically devaluing their currency which, in addition to the geopolitical problems, would create havoc for the currencies of the rest of the world - likely forcing them to massively inflate their own currencies to match. Perhaps deficits will simply rise to the point where even with low interest rates some nations will be unable to make the interest payments on their debt. This would force sovereign loans be either restructured -leading to contagion in nations around the world - or be monetized.

Even if for a moment we put aside these scenarios, some things are now inevitable. Budget deficits cannot continue *indefinitely* in a world with declining working-age populations. At least not without simply printing the money. Governments make up the deficits in their budgets by borrowing it. At the moment nations are borrowing from the Baby Boomer generation - when they retire this simply will not work anymore. Even ignoring the very real potential for crisis, just the shock of balanced budgets would be immense. In 2018 the U.S. ran a budget deficit of more than a trillion dollars.

By 2049, the budget deficit in the U.S. is projected to be double today's deficit as a percent of GDP. If *somehow* crisis was avoided, and the only consequence was the nations were - at some point in the future - forced to balance their budgets, think of the havoc this would cause. A balanced budget today would mean cutting almost the equivalent of the entire defense budget *and* all spending for Medicare. And again, the deficit is only projected to widen in the coming decades.

This is the path that the world is on with the ageing of the population and the decline of working age replacements. As the massive Baby Boomer generation transitions from being net buyers of stocks and bonds to net sellers, and from taxpayers to pensioners, national budgets and ultimately national debts will come under increasing pressure. A loss of confidence in sovereign debt would leave nations with impossible choices and risks economic downturns on the scale of the Great Depression or Weimar Republic style inflation. As indicators have begun to warn that the global economy may soon enter a recession, the risks to sovereign debt will rise tremendously.

Debt Forgiveness

One way or another the debt will need to be eliminated. It cannot realistically be repaid. Many of those who understand the unsustainable nature of the debt believe that an alternative to monetization or default would be a Jubilee. Some analysts believe that because central banks now own so much of the world's sovereign debt that they can simply forgive this debt. Indeed, the BOJ now owns almost 70% of all outstanding JGBs. Up from 7.8% in 2008. Central banks QE programs have lead them to be the owner of a substantial amount of government's debt in many nations. What would stop the BOJ from announcing that it would convert this debt into 100 year bonds that paid no interest. Effectively erasing this debt.

But effectively, Japan and other nations has already done this. Interest on the debt held by central banks are returned to the treasury. But that does not mean that the debt has been erased. If you recall the process of QE, the debt held by the central bank was purchased from the banking system and banks were given cash which is held at the central bank and these reserves at the central bank typically pay an interest to the banks. The Federal Reserve, for example, currently pays 2.35% on reserves held with the central bank just a tenth of a percent less than the current 10 year yield, QE pushes down interest rates, but it does not eliminate the effective burden completely. To erase the actual impact of the accumulated debt, central banks would need to stop paying interest on the reserves held at banks and prevent the money from flowing into the economy by freezing these assets.

Even if those hurdles could be overcome and the debt erased, the real problem would still remain. Pension systems are still reliant on their holdings of sovereign debt, and foreign investors hold sovereign debt. If *that* debt is purchased with cash from the central bank it would be true monetization. Moreover, who will purchase sovereign debt in the future to maintain nation's budget deficits? Not pension funds, they will become net sellers. Not banks – at least not if you've stopped paying them interest on reserves and have frozen their reserves. Not to mention that banks have finite assets.

Ultimately, there are only two choices. Default on the debt and endure a deflationary spiral, or monetize the debt and endure significant inflation.

Notes

SLEEPING ON A VOLCANO

1   Craig, Susanne. "In Former C.E.O.'s Words, the Last Days of Lehman Brothers." *The New York Times*, February 14, 2011. Accessed April 22, 2019. https://dealbook.nytimes.com/2011/02/14/a-different-side-to-dick-fuld/.

2   Kessler, Andy. "Lehman: Political and Personal." *The Wall Street Journal*, September 9, 2018. Accessed April 22, 2019. https://www.wsj.com/articles/lehman-political-and-personal-1536524375?ns=prod/accounts-wsj.

3   Shea, James. "Sen. Burr Speaks on Economy." *Hendersonville Times-News*, April 14, 2009. https://www.blueridgenow.com/news/20090414/sen-burr-speaks-on-economy.

4   Amadeo, Kimberly. "Where Were You the Day the U.S. Economy Nearly Collapsed?" *The Balance*, November 15, 2018. https://www.thebalance.com/reserve-primary-fund-3305671.

5   Solomon, Deborah, Liz Rappaport, Damian Paletta, and Jon Hilsenrath. "Shock Forced Paulson's Hand: A Black Wednesday on Credit Markets; 'Heaven Help Us All'." *The Wall Street Journal*, September 20, 2008. https://www.wsj.com/articles/SB122186563104158747?ns=prod/accounts-wsj.

6   Solomon, Deborah, Liz Rappaport, Damian Paletta, and Jon Hilsenrath. "Shock Forced Paulson's Hand: A Black Wednesday on Credit Markets; 'Heaven Help Us All'." *The Wall Street Journal*, September 20, 2008. https://www.wsj.com/articles/SB122186563104158747?ns=prod/accounts-wsj.

7   M. HERSZENHORN, DAVID, CARL HULSE, and SHERYL GAY STOLBERG. "Talks Implode During a Day of Chaos; Fate of Bailout Plan Remains Unresolved." *The New York Times*, September 25, 2008. https://www.nytimes.com/2008/09/26/business/26bailout.html.

8   Foley, Stephen. "Crash of a Titan: The inside Story of the Fall of Lehman Brothers." *The Independent*, September 7, 2009. https://www.independent.co.uk/news/business/analysis-and-features/crash-of-a-titan-the-inside-story-of-the-fall-of-lehman-brothers-1782714.html.

9   Ivry, Bob, Bradley Keoun, and Phil Kuntz. "Secret Fed Loans Gave Banks $13 Billion Undisclosed to Congress." *Bloomberg: Markets Magazine*, November 27,

2011. https://www.bloomberg.com/news/articles/2011-11-28/secret-fed-loans-undisclosed-to-congress-gave-banks-13-billion-in-income.

10 Barrett, Paul. "Prophet-Making." The New York Times. June 25, 2010. Accessed June 06, 2019. https://www.nytimes.com/2010/06/27/books/review/Barrett-t.html.

11 Balzli, Beat, and Michaela Schießl. "Global Banking Economist Warned of Coming Crisis." *Spiegel*, July 8, 2009. https://www.spiegel.de/international/business/the-man-nobody-wanted-to-hear-global-banking-economist-warned-of-coming-crisis-a-635051.html.

12 Rushton, Katherine. "Ken Rogoff Warns China Is next Bubble to Burst." The Telegraph. May 17, 2014. Accessed June 04, 2019. https://www.telegraph.co.uk/-finance/economics/10838674/Ken-Rogoff-warns-China-is-next-bubble-to-burst.html.

13 Burry, Michael J. "I Saw the Crisis Coming. Why Didn't the Fed?" The New York Times. April 03, 2010. Accessed June 04, 2019. https://www.nytimes.com/2010/04/04/opinion/04burry.html.

14 "Letter From The Mortgage Insurance Companies of America." Suzanne C. Hutchinson to Board of Governors of the Federal Reserve System. March 29, 2006.
   https://www.fdic.gov/regulations/laws/federal/2005/05c45guide.pdf

15 Evans-Pritchard, Ambrose. "World Finance Now More Dangerous than in 2008, Warns Central Bank Guru." *The Telegraph*, January 22, 2008. https://www.telegraph.co.uk/business/2018/01/22/world-finance-now-dangerous-2008-warns-central-bank-guru/.

16 Evans-Pritchard, Ambrose. "World Finance Now More Dangerous than in 2008, Warns Central Bank Guru." *The Telegraph*, January 22, 2008. https://www.telegraph.co.uk/business/2018/01/22/world-finance-now-dangerous-2008-warns-central-bank-guru/.

17 Furness, Virginia. "Debt Bubbles Could Burst as Economies Slow, Threatening Stability,..." Reuters. May 16, 2019. Accessed June 14, 2019. https://www.reuters.com/article/us-capital-markets-bonds-idUSKCN1SM2GU.

18 Inman, Phillip. "World Economy at Risk of Another Financial Crash, Says IMF." The Guardian. October 03, 2018. Accessed August 11, 2019. https://www.theguardian.com/business/2018/oct/03/world-economy-at-risk-of-another-financial-crash-says-imf.

THE DEMOGRAPHIC UPHEAVAL AND ITS ECONOMIC IMPACT

19 Singh, Salil. "Norman Borlaug: A Billion Lives Saved." AgBioWorld. 2011. http://www.agbioworld.org/biotech-info/topics/borlaug/special.html.

NOTES

20 The United Nations. "Report of the Second World Assembly on Ageing." Proceedings of Second World Assembly on Ageing, Madrid. April 2002. http://www.cpahq.org/cpahq/cpadocs/UN Report of the Second World Assembly on Ageing.pdf.

21 Galor, Oded. *Fertility and Growth.* Department of Economics, Brown University. April 8, 2012. https://www.copenhagenconsensus.com/sites/default/files/populationgrowth_perspectivepapergalor_0.pdf.

22 Kotecki, Peter. "10 Countries at Risk of Becoming Demographic Time Bombs." *Business Insider*, August 8, 2018. https://www.businessinsider.com/10-countries-at-risk-of-becoming-demographic-time-bombs-2018-8.

23 The United Nations. "Report of the Second World Assembly on Ageing." Proceedings of Second World Assembly on Ageing, Madrid. April 2002. http://www.cpahq.org/cpahq/cpadocs/UN Report of the Second World Assembly on Ageing.pdf.

24 University of California - Los Angeles. "People entering their 60s may have more disabilities today than in prior generations." ScienceDaily. www.sciencedaily.com/releases/2009/11/091112162832.htm (accessed April 22, 2019).

25 Peterson, Peter G. "Gray Dawn: The Global Aging Crisis." *Foreign Affairs* 78, no. 1 (1999): 42. doi:10.2307/20020238.

26 Goodhart, Charles. "Demographics Will Reverse Three Multi-decade Global Trends." *BIS Working Papers*, August 7, 2017. https://www.bis.org/publ/-work656.htm.

27 UNFPA. "Demographic Dividend." Accessed May 31, 2019. https://botswana.unfpa.org/en/topics/demographic-dividend-0.

28 Mason, Andrew. "Demographic Transition and Demographic Dividends in Developed and Developing Countries." January 2005. https://www.un.org/en/development/desa/population/events/pdf/expert/9/mason.pdf.

29 Bloom, David E., and Jocelyn E. Finlay. "Demographic Change and Economic Growth in Asia." September 2008. https://cdn1.sph.harvard.edu/wp-content/uploads/sites/1288/2013/10/PGDA_WP_41.pdf.

30 Peterson, Peter G. "Gray Dawn: The Global Aging Crisis." *Foreign Affairs* 78, no. 1 (1999): 42. doi:10.2307/20020238.

31 Kim, Jinill. "The Effects of Demographic Change on GDP Growth in OECD Economies." *IFDP Notes: Washington: Board of Governors of the Federal Reserve System*, September 28, 2016. doi:https://doi.org/10.17016/2573-2129.22.

32 Goodhart, Charles. "Demographics Will Reverse Three Multi-decade Global Trends." *BIS Working Papers*, August 7, 2017. https://www.bis.org/publ/-work656.htm.

33 Schieber, Sylvester, and John Shoven. "The Consequences of Population Aging on Private Pension Fund Saving and Asset Markets." 1994. doi:10.3386/w4665.

34 Takats, Elod. "Ageing and Asset Prices." *SSRN Electronic Journal*, 2010. doi:10.2139/ssrn.1672638.

35 McMahon, Jeff. "The World Economy Is A Pyramid Scheme, Steven Chu Says." Forbes. April 05, 2019. Accessed June 03, 2019. https://www.forbes.com/sites/jeffmcmahon/2019/04/05/the-world-economy-is-a-pyramid-scheme-steven-chu-says/#6680dcb14f17.

THE RISE AND FALL OF THE BRETTON WOODS SYSTEM

36 Sterilizing gold flows means that the central banks - for example the United States and France - sterilized their gold inflows - as gold flowed into the United States and France, rather than allowing it to increase the money supply, they worked to offset the gold by contracting credit to the banks. Preventing the inflation that it would have otherwise. See *The Lords of Finance: The Bankers Who Broke the World* by Liaquat Ahamed

37 explain

38 See for example: Eichengreen, Barry, and Kris J. Mitchener. "The Great Depression As A Credit Boom Gone Wrong." *Research in Economic History*, September 2003, 183-237. doi:10.1016/s0363-3268(04)22004-3.

39 Ahamed, Liaquat. *Lords of Finance.* London: Windmill Books, 2010.

40 Hull, Cordell. *The Memoirs of Cordell Hull.* London: Hodder & Stoughton, 1948.

41 Steil, Benn. *The Battle of Bretton Woods: John Maynard Keynes, Harry Dexter White, and the Making of a New World Order.* Princeton: Princeton University Press, 2013.

42 Shelton, Judy. *Money Meltdown: Restoring Order to the Global Currency System.* New York: Free Press, 1994.

43 Eichengreen, Barry J. *Exorbitant Privilege: The Rise and Fall of the Dollar and the Future of the International Monetary System.* Oxford: Oxford University Press, 2012.

44 "Money Matters, an IMF Exhibit — The Importance of Global Cooperation, System in Crisis (1959-1971), Part 4 of 7." International Monetary Fund. Accessed June 03, 2019. https://www.imf.org/external/np/exr/center/mm/eng/mm_sc_03.htm.

45 Reinbold, Brian, and Yi Wen. "Understanding the Roots of the U.S. Trade Deficit." St. Louis Fed. October 12, 2018. Accessed June 03, 2019. https://www.stlouisfed.org/publications/regional-economist/third-quarter-2018/understanding-roots-trade-deficit.

46 Mundell, Robert. "The Case for a World Currency." *Journal of Policy Modeling* 34, no. 4 (2012): 568-78. doi:10.1016/j.jpolmod.2012.05.011.

47 Griswold, Daniel. "America's Misunderstood Trade Deficit." Cato Institute. July

NOTES

22, 1998. Accessed August 16, 2019. https://www.cato.org/publications/congressional-testimony/americas-misunderstood-trade-deficit.

48 Petteis, Michael. *The Great Rebalancing.* Princeton University Press, 2015.

49 Petteis, Michael. *The Great Rebalancing.* Princeton University Press, 2015.

50 King, Mervyn. "World Trade and Exchange Rates: From the Pax Americana to a Multilateral New Order." Speech, Peterson Institute of International Economics, Washington DC. May 16, 2017. https://www.piie.com/system/files/documents/king20170516.pdf.

51 Petteis, Michael. *The Great Rebalancing.* Princeton University Press, 2015.

52 Benko, Ralph. "Keynes and Copernicus: The Debasement Of Money Overthrows The Social Order And Governments." Forbes. December 23, 2013. Accessed June 03, 2019. https://www.forbes.com/sites/ralphbenko/2013/12/23/keynes-and-copernicus-the-debasement-of-money-overthrows-the-social-order-and-governments/#1f80b4d25fc4.

53 Herradi, Mehdi El, and Aurélien Leroy. "Monetary Policy and the Top One Percent: Evidence from a Century of Modern Economic History." *SSRN Electronic Journal*, 2019. doi:10.2139/ssrn.3379740.

54 Herradi, Mehdi El, and Aurélien Leroy. "Monetary Policy and the Top One Percent: Evidence from a Century of Modern Economic History." *SSRN Electronic Journal*, 2019. doi:10.2139/ssrn.3379740.

THE ACCUMULATION OF DEBT AND QUANTITATIVE EASING

55 Peter G. Peterson Foundation. "CBO Warns: Fiscal Outlook Remains Unsustainable." Peter G. Peterson Foundation. 2017. Accessed August 11, 2019. https://www.pgpf.org/analysis/2017/03/as-policymakers-consider-changes-cbo-warns-fiscal-outlook-remains-unsustainable.

56 "America's Fiscal Future - Current Fiscal Conditions." U.S. Government Accountability Office (U.S. GAO). Accessed August 11, 2019. https://www.gao.gov/americas_fiscal_future?t=current_fiscal_conditions.

57 Wallace, Tim. "Ageing Population Puts UK on Track for 'unsustainable' Surge in Debt, Warns OBR." The Telegraph. July 17, 2018. Accessed August 11, 2019. https://www.telegraph.co.uk/business/2018/07/17/ageing-population-puts-uk-track-unsustainable-surge-debt/.

58 Hoshi, Takeo, and Takatoshi Ito. "Defying Gravity: Can Japanese Sovereign Debt Continue to Increase without a Crisis?" *Economic Policy* 29, no. 77 (2014): 5-44. doi:10.1111/1468-0327.12023.

59 Rogoff, Kenneth S., and Carmen M. Reinhart. *This Time Is Different.*

60 OECD. *OECD Sovereign Borrowing Outlook 2018.* Report. 2018. http://www.oecd.org

/daf/fin/public-debt/Sovereign-Borrowing-Outlook-in-OECD-Countries-2018.pdf.

61 Carvalho, Carlos, Andrea Ferrero, and Fernanda Nechio. "Demographics and Real Interest Rates: Inspecting the Mechanism." *SSRN Electronic Journal*, 2016. doi:10.2139/ssrn.2713443.

62 "Jacksonville University Finance Discussion." In *Jacksonville University Finance Discussion*. C-SPAN. November 5, 2010.
   Federal Reserve Chairman Ben Bernanke

63 Gagnon, Joseph, Matthew Raskin, Julie Remache, and Brian Sack. "The Financial Market Effects of the Federal Reserve's Large-Scale Asset Purchases." *International Journal of Central Banking*, March 2011. https://www.ijcb.org/journal/ijcb11q1a1.pdf.

64 Bernanke, Ben S. "The Latest from the Bank of Japan." Brookings. September 26, 2016. Accessed August 11, 2019. https://www.brookings.edu/blog/ben-bernanke/2016/09/21/the-latest-from-the-bank-of-japan/.

65 Moore, Elaine. "When Will the ECB Run out of Bonds to Buy?" Financial Times. September 08, 2016. Accessed August 11, 2019. https://www.ft.com/content/4857b806-7436-11e6-b60a-de4532d5ea35.

66 "Limits in Terms of Eligible Collateral and Policy Risks of an Extension of the ECB's Quantitative Easing Programme." Kiel Institute - Top Research in Global Economic Affairs. Accessed August 11, 2019. https://www.ifw-kiel.de/publications/policy-papers
   /analysen-fuer-das-europaeische-parlament/limits-in-terms-of-eligible-collateral-and-policy-risks-of-an-extension-of-the-ecbs-quantitative-easing-
   programme-10093/.

67 "Limits in Terms of Eligible Collateral and Policy Risks of an Extension of the ECB's Quantitative Easing Programme." Kiel Institute - Top Research in Global Economic Affairs. Accessed August 11, 2019. https://www.ifw-kiel.de/publications/policy-papers/analysen-fuer-das-europaeische-parlament/limits-in-terms-of-eligible-collateral-and-policy-risks-of-an-extension-of-the-ecbs-quantitative-easing-programme-10093/.

68 Caballero, Ricardo, Takeo Hoshi, and Anil Kashyap. "Zombie Lending and Depressed Restructuring in Japan." 2006. doi:10.3386/w12129.

69 Banerjee, Ryan. "The Rise of Zombie Firms: Causes and Consequences." *BIS Quarterly Review*, September 2018. https://www.bis.org/publ/qtrpdf/r_qt1809g.pdf.

70 Lacalle, Daniel. "The Rise of Zombie Companies, And Why It Matters To You." July 20, 2017. Accessed August 11, 2019. https://www.dlacalle.com/en/the-rise-of-zombie-companies-and-why-it-matters-to-you/.

NOTES

71 The Pew Charitable Trusts. "State Public Pension Funds' Investment Practices and Performance: 2016 Data Update." The Pew Charitable Trusts. September 26, 2018. Accessed August 12, 2019. https://www.pewtrusts.org/en/research-and-analysis/issue-briefs/2018/09/state-public-pension-funds—investment-practices-and—performance-2016-data-update.

72 The Pew Charitable Trusts. "State Public Pension Funds' Investment Practices and Performance: 2016 Data Update." The Pew Charitable Trusts. September 26, 2018. Accessed August 12, 2019. https://www.pewtrusts.org/en/research-and-analysis/issue-briefs/2018/09/state-public-pension-funds—investment-practices-and—performance-2016-data-update.

73 Williams, Jonathan, Christine Smith, Thurston Powers, and Bob Williams. "Unaccountable and Unaffordable 2018." American Legislative Exchange Council. March 20, 2019. Accessed August 12, 2019. https://www.alec.org/publication/unaccountable-and-unaffordable-2018/.

JAPAN

74 Hoshi, Takeo, and Takatoshi Ito. "Defying Gravity: Can Japanese Sovereign Debt Continue to Increase without a Crisis?" *Economic Policy* 29, no. 77 (2014): 5-44. doi:10.1111/1468-0327.12023.

75 Miyazaki, Tomomi, and Kazuki Onji. "The Sustainability of Japan's Government Debt: A Review." *Theoretical Economics Letters* 07, no. 06 (2017): 1632-645. doi:10.4236/tel.2017.76110.

76 Doi, Takero, Takeo Hoshi, and Tatsuyoshi Okimoto. "Japanese Government Debt and Sustainability of Fiscal Policy." 2011. doi:10.3386/w17305.

77 Undefined, Undefined Undefined. *The Economic Effects of Aging in the United States and Japan.* By Michael D. Hurd and Naohiro Yashiro. Chicago: University of Chicago Press, 1997.

78 Lam, Raphael W., and Kiichi Tokuoka. "Assessing the Risks to the Japanese Government Bond (JGB) Market." *IMF Working Papers* 11, no. 292 (2011): I. doi:10.5089/9781463927264.001.

ITALY AND EUROPE

79 "German Unification: "Thatcher Told Gorbachev Britain Did Not Want German Unification" (documents from Gorbachev Archive) ["Britain & Western Europe Are Not Interested in the Unification of Germany"]." Margaret Thatcher Foundation. Accessed August 14, 2019. https://www.margaretthatcher.org/document/112006.

80 Volkery, Carsten. "The Iron Lady's Views on German Reunification: 'The Germans Are Back!' - SPIEGEL ONLINE - International." SPIEGEL ONLINE. September 11, 2009. Accessed August 14, 2019. https://www.spiegel.de/international/europe/the-

iron-lady-s-views-on-german-reunification-the-germans-are-back-a-648364.html.

81 Marshall, Tim. *Prisoners of Geography: Ten Maps That Explain Everything about the World*. New York: Scribner Book Company, 2016.

82 Kaldor, Nicholas. *Further Essays on Applied Economics*. London: Duckworth, 1978.

83 Estenssoro, Amalia. "European Sovereign Debt Remains Largely a European Problem." Federal Reserve Bank of St. Louis. June 28, 2016. Accessed August 14, 2019. https://www.stlouisfed.org/publications/regional-economist/october-2010/european-sovereign-debt-remains-largely-a-european-problem.

84 Flassbeck, Heiner. "Germany's Trade Surplus: Causes and Effects." *American Affairs* 1, no. 3 (Fall 2017). https://americanaffairsjournal.org/2017/08/germanys-trade-surplus-causes-effects/.

85 Conway, Edmund, and Daily Telegraph. "Europe Is Headed for a Meltdown." RealClearPolitics. May 27, 2010. Accessed August 14, 2019. https://www.realclearworld.com/2010/05/27/europe_is_headed_for_a_meltdown_111937.html#!

86 Varoufakis, Yanis. *Adults In The Room*. Random House UK, 2018.

87 Walker, Marcus. "Greek Debt Crisis: Germany Flexes Its Muscles in Talks With Bailout Ultimatum." The Wall Street Journal. July 13, 2015. Accessed August 14, 2019. https://www.wsj.com/articles/greek-debt-crisis-europe-pushes-athens-to-brink-with-bailout-ultimatum-1436745387.

88

89 Horowitz, Jason. "Italy's Toying With a 'Mini-BOT' Worries E.U. and Investors." The New York Times. The New York Times, June 13, 2019. https://www.nytimes.com/2019/06/13/world/europe/italy-mini-bot-eu-brussels-debt.html.

90 Evans-Pritchard, Ambrose. "Italy to Activate Its 'parallel Currency' in Defiant Riposte to EU Ultimatum." The Telegraph. May 29, 2019. Accessed June 17, 2019. https://www.telegraph.co.uk/business/2019/05/29/epic-clash-building-italys-triumphant-salvini-brussels/.

91 Jones, Gavin. "Italy's Dual Currency Schemes May Be Long Road to Euro Exit." Reuters. September 08, 2017. Accessed June 17, 2019. https://www.reuters.com/article /us-italy-euro-analysis-idUSKCN1BJ20F.

92 Jones, Gavin. "Italy's Dual Currency Schemes May Be Long Road to Euro Exit." Reuters. September 08, 2017. Accessed June 17, 2019. https://www.reuters.com/article/us-italy-euro-analysis-idUSKCN1BJ20F.

93 "Italy's Scary Parallel Currency Threat." Yahoo! Finance. June 07, 2019. Accessed June 17, 2019. https://finance.yahoo.com/news/italy-apos-scary-parallel-currency-063022000.html.

NOTES

94 Lachman, Desmond. "Italy Is Damned If It Does, Damned If It Doesn't." TheHill. June 04, 2019. Accessed August 14, 2019. https://thehill.com/opinion/finance/446886-italy-is-damned-if-it-does-damned-if-it-doesnt.

CHINA

95 People's Daily Online. "The Xiaogang Village Story." People's Daily Online. November 11, 2008. Accessed August 14, 2019. http://en.people.cn/90002/95607/6531490.html.

96 Scott, Robet E. "Costly Trade With China: Millions of U.S. Jobs Displaced with Net Job Loss in Every State." Economic Policy Institute. October 9, 2007. Accessed August 14, 2019. https://www.epi.org/publication/bp188/.

97 Chen, Wei, Chang-Tai Hsieh Xilu Chen, Zheng, and Michael. "A Forensic Examination of China's National Accounts." Brookings. March 15, 2019. Accessed August 14, 2019. https://www.brookings.edu/bpea-articles/a-forensic-examination-of-chinas-national-accounts/.

98 Eberstadt, Nicholas. "With Great Demographics Comes Great Power." *Foreign Affaris*, July/August 2019.

99 Anderlini, Jamil. "China's Economy Is Addicted to Debt." Financial Times. August 15, 2017. Accessed August 14, 2019. https://www.ft.com/content/293f8f22-8112-11e7-94e2-c5b903247afd.

100 Lam, Eric. "China May Have $5.8 Trillion in Hidden Debt With 'Titanic' Risks." Bloomberg.com. October 16, 2018. Accessed August 14, 2019. https://www.bloomberg.com/news/articles/2018-10-16/china-may-have-5-8-trillion-in-hidden-debt-with-titanic-risks.

101 Persinos, John. "China's $5 Trillion in Toxic Bank Debt Is about to Collapse — Here's How to Profit." TheStreet. February 05, 2016. Accessed August 14, 2019. https://www.thestreet.com/story/13449814/1/china-s-5-trillion-in-toxic-bank-debt-is-about-to-collapse-here-s-how-to-profit.html.

102 Bird, Mike. "China's Banks Are Running Out of Dollars." The Wall Street Journal. April 23, 2019. Accessed August 14, 2019. https://www.wsj.com/articles/chinas-banks-are-running-out-of-dollars-11556012442.

103 Balding, Christopher. "The Dollar Dictates China's Need for a Trade Deal." Bloomberg.com. May 8, 2019. Accessed August 14, 2019. https://www.bloomberg.com/opinion/articles/2019-05-09/china-s-need-for-u-s-dollars-dictates-its-trade-stance.

104 "The Global Asset Bubble Will Burst, Just Not in 2018." South China Morning Post. January 06, 2018. Accessed August 14, 2019. https://www.scmp.com/comment/insight-opinion/article/2126952/global-asset-bubble-will-burst-only-question-when-and-how.

105 Tan, Huileng. "China Could Export a Recession to Everyone Else, Says Ex-IMF Chief Economist Kenneth Rogoff." CNBC. July 06, 2017. Accessed August 14, 2019. https://www.cnbc.com/2017/07/05/ex-imf-chief-economist-rogoff-china-could-export-a-recession.html.

THE UNITED STATES

106 Krauss, Clifford. "Oil Boom Gives the U.S. a New Edge in Energy and Diplomacy." The New York Times. January 28, 2018. Accessed August 13, 2019. https://www.nytimes.com/2018/01/28/business/energy-environment/oil-boom.html.

107 Donnan, Shawn. "US Says China WTO Membership Was a Mistake." Financial Times. January 19, 2018. Accessed August 13, 2019. https://www.ft.com/content/edb346ec-fd3a-11e7-9b32-d7d59aace167.

108 Zeihan, Peter. *The Accidental Superpower: The next Generation of American Preeminence and the Coming Global Disorder*. New York: Twelve, 2016.

109 Gallup, Inc. "Most Important Problem." Gallup.com. Accessed June 03, 2019. https://news.gallup.com/poll/1675/most-important-problem.aspx.

110 Gallup, Inc. "Budget Rises as Most Important Problem to Highest Since '96." Gallup.com. Accessed June 03, 2019. https://news.gallup.com/poll/147086/budget-rises-most-important-problem-highest.aspx.

111 "Military Spending around the World Is Booming." The Economist. April 28, 2019. Accessed June 03, 2019. https://www.economist.com/international/2019/04/28/military-spending-around-the-world-is-booming.

112 "Should Fair-Value Accounting Be Used to Measure the Cost of Federal Credit Programs?" Congressional Budget Office. Accessed June 03, 2019. https://www.cbo.gov/publication/43035.

113 "The Case for Fair-Value Accounting." National Affairs. Accessed June 03, 2019. https://www.nationalaffairs.com/publications/detail/the-case-for-fair-value-accounting.

114 Reinhart, Carmen M., and Kenneth S. Rogoff. *This Time Is Different: Eight Centuries of Financial Folly*. Princeton, NJ: Princeton University Press, 2011.

THE SOVEREIGN DEBT CRISIS

115 Minsky, Hyman P. *Can "it" Happen Again?: Essays on Instability and Finance*. Abingdon, Oxon: Routledge, 2016.

116 Mises, Ludwig Von. *Economic Calculation in the Socialist Commonwealth*. Auburn, Ala.: Ludwig Von Mises Institute, Auburn University, 2008.

117 Eichengreen, Barry, and Kris J. Mitchener. "The Great Depression As A Credit Boom Gone Wrong." *Research in Economic History*, August 2003, 183-237. doi:10.1016/s0363-3268(04)22004-3.

NOTES

118 Richardson, Gary. "Banking Panics of 1930-31." Federal Reserve History. November 22, 2013. Accessed June 03, 2019. https://www.federalreservehistory.org/essays/banking_panics_1930_31.

119 Mason, John. "Why Larry Summers Is Completely Wrong on How to Fix the U.S. Economy." TheStreet. February 20, 2016. Accessed June 03, 2019. https://www.thestreet.com/story/13464339/1/why-larry-summers-is-completely-wrong-on-how-to-fix-the-u-s-economy.html.

120 Blomberg Editorial Board. "Breaking the Sovereign-Debt Doom Loop." Bloomberg.com. December 26, 2017. https://www.bloomberg.com/opinion/articles/2017-12-26/breaking-the-sovereign-debt-doom-loop.

121 Cecchetti, Stephen G., Madhusudan S. Mohanty, and Fabrizio Zampolli. "The Future of Public Debt: Prospects and Implications." *SSRN Electronic Journal*, 2010. doi:10.2139/ssrn.1599421.

122 Evans-Pritchard, Ambrose. "Central Bank Prophet Fears QE Warfare Pushing World Financial System out of Control." The Telegraph. January 20, 2015. Accessed June 12, 2019. https://www.telegraph.co.uk/finance/economics/11358316/Central-bank-prophet-fears-QE-warfare-pushing-world-financial-system-out-of-control.html.

Printed in Great Britain
by Amazon